Reproducible Activities

Standards-Based Math

Grades 1-2

By
Melissa J. Owen

Cover Design by
Jeff Van Kanegan

Inside Illustrations by
Rebecca Waske

Published by Instructional Fair • TS Denison
an imprint of

About the Author

M. J. Owen holds a master's degree in education from the University of Houston. She has worked in public schools in Yarmouth, Maine, and rural and central Texas. She is at home in Austin, Texas where she enjoys writing, running, and traveling.

M. J. dedicates this book to her daughter, Ellie, with the hopes she always has fun (and success) in math.

Credits

Author: Melissa J. Owen
Cover Design: Jeff Van Kanegan
Inside Illustration: Rebecca Waske
Project Director/Editor: Sara Bierling
Editors: Mary Rose Hassinger, Tiauna Harris
Graphic Layout: Tracy L. Wesorick

McGraw-Hill Children's Publishing

A Division of The McGraw-Hill Companies

Published by Instructional Fair • TS Denison
An imprint of McGraw-Hill Children's Publishing
Copyright © 2002 McGraw-Hill Children's Publishing

Limited Reproduction Permission: Permission to duplicate these materials is limited to the person for whom they are purchased. Reproduction for an entire school or school district is unlawful and strictly prohibited.

Send all inquiries to:
McGraw-Hill Children's Publishing
3195 Wilson Drive NW
Grand Rapids, Michigan 49544

All Rights Reserved • Printed in the United States of America

Standards-Based Math—grades 1–2
ISBN: 0-7424-0214-2

1 2 3 4 5 6 7 8 9 07 06 05 04 03 02

Introduction

Standards-Based Math is based on the most recent standards from the National Council of Teachers of Mathematics. The proposed ten NCTM standards, which are a description of what an existing curriculum should enable students to know and do, can be divided into two categories. Content standards are those five that state the content (skills) that students should be taught and should learn. These are Number and Operations, Algebra, Geometry, Measurement, and Data Analysis and Probability. The Process standards, which guide instructors in facilitating lessons that enable students to acquire knowledge and achieve success, are the second set of standards. These standards are Problem Solving, Reasoning and Proof, Communication, Connections, and Representation.

For the purpose of this book, the Content standards have been highlighted, and each page is labeled with its appropriate standard and skill. The Process standards are woven into the exercises on each page.

The following are simplified descriptions of the Content standards.

Number and Operations
1. Understands numbers, number representation, relationships, and systems
2. Comprehends operations and their relatedness
3. Fluent in computation and estimation

Algebra
1. Recognizes relationships, functions, and patterns in numbers
2. Uses algebraic symbols
3. Shows similarities in math equations using models
4. Recognizes and evaluates change

Geometry
1. Has knowledge of two- and three-dimensional objects
2. Uses coordinate geometry
3. Recognizes symmetry and transformation
4. Solves problems using geometry

Measurement
1. Understands measurement
2. Uses a variety of methods to show measurement

Data Analysis and Probability
1. Understands the necessity of data
2. Utilizes statistics
3. Makes inferences and predictions based on data
4. Comprehends the basic principles of probability

Table of Contents

Number and Operations

- Mix and Match Socks (counting) 6
- Trace This (writing numbers)..........................7
- Ones and Tens (place value) 8
- A House Full of Place Value (place value) ..9
- Mystery Numbers (place value) 10
- Finding the "In Between" (relative position) 11
- Teddy Bears (number words and numerals) 12
- Understanding Fractions (fractions) 13
- Cupcake Fractions (fractions).....................14
- The Whole Team (fractions).......................15
- Cut and Paste Fractions (fractions) 16
- Show What You Know! (fractions) 17
- Who Is Bigger? (greater than 10) 18
- Hungry Hippos (comparison).......................19
- Compute with Caroline (addition) 20
- Rover's Bones (addition) 21
- Hot Air Balloons (addition).........................22
- Adding It Up (addition) 23
- Delicious Word Problems (addition) 24
- The Answer Is... (missing addend) 25
- Sliding Through Subtraction (subtraction) 26
- Bouncing Balls (subtraction) 27
- Subtraction Practice (subtraction) 28
- A Bunch of Bubbles (2-digit subtraction) 29
- Worldly Word Problems (subtraction)..........30
- Coins (money)...31
- Maggie's Wallet (money) 32
- A Day at the Mall (money) 33
- Louie's Loose Change (money) 34
- Coloring Coins (money) 35
- Solve with Words (addition and subtraction) 36
- Find the Sign (meaning of operations)........37
- Spending and Earning Money (meaning of operations) 38
- Practice, Practice (2-digit addition with regrouping)...............39
- Mousetrap (2-digit addition with regrouping)...............40
- Ice Cream Scoops (2-digit addition with regrouping)...............41
- Adding Along (2-digit addition with regrouping)...............42
- Hats Off to You! (2-digit subtraction with regrouping) 43
- Subtraction Maze (2-digit subtraction with regrouping) 44
- A Piece of Cake! (2-digit subtraction with regrouping) 45
- Nutty Solutions (2-digit subtraction with regrouping) 46
- A Cool Drink (2-digit subtraction with regrouping) 47
- Curtains Up on Addition (3-digit addition with regrouping)...............48
- You're Bright (3-digit addition with regrouping)...............49
- Sweet Treats (3-digit addition with regrouping)...............50
- Up and Away (3-digit addition with regrouping)...............51
- Big Number on Top! (3-digit subtraction with regrouping) 52
- A Letter Maze (3-digit subtraction with regrouping) 53
- Tempting Snacks (3-digit subtraction with regrouping) 54
- Secret Code (3-digit subtraction with regrouping) 55
- Groups (multiplication/grouping) 56
- Cookies, Cookies, Cookies! (multiplication/grouping)..............................57

Sharing with Friends (division/sharing) 58
Birthday Party Guests (division/sharing) 59

Algebra

Small to Large (sort and classify) 60
Sort It Out (sort and classify) 61
Skipping Stones (skip counting)....................62
Number Patterns (patterning) 63
Square, Square, Circle (patterning) 64
Sizable Patterns (patterning) 65
Pasting Patterns (patterning)......................66
Candy Store Patterns (patterning) 67
It's the Opposite (commutativity)................68
Growing Up (qualitative change) 69
Solving Symbols
(representing with symbols) 70

Geometry

Flat Shapes (shapes)....................................71
Flying High (shapes) 72
Discovering Details! (shapes).......................73
Three-Dimensional Shapes (3-D shapes) 74
In the Air (shapes) 75
Shapes Take Off (shapes) 76
Who Am I? (shapes) 77
Symmetry (symmetry) 78
Find My Match (symmetry) 79
Simply Symmetrical (symmetry) 80
Mirror, Mirror, on the Wall (symmetry).........81
Lines of Symmetry (symmetry) 82
Double Take (flips) 83
Shapes Around Us (shapes).......................84
Symmetry Review (symmetry) 85

Measurement

Counting Caterpillars
(nonstandard measurement) 86
Paper Clip It
(nonstandard measurement) 87
Looking for Length (inches) 88

Taking the Measurement (inches) 89
Traveling with Squirrel
(centimeters/meters) 90
Munching (centimeters).............................91
What Is the Wacky Weight? (pounds) 92
Basket Weight (kilograms) 93
All About Area (area)................................94
Measurement Review
(length, weight, area) 95
At the Race (cups, pints, quarts) 96
Telling Time (time) 97
The Zookeeper's Watch (time) 98
Don't Be Late! (time) 99
Watching the Time (time) 100
Time Puzzlers (time) 101
Vacation Time! (minutes, hours, days,
weeks, months) 102

Data Analysis and Probability

Taking Notes (tally chart) 103
On the Chart (bar graph) 104
Pizza Pictograph (pictograph) 105
Guessable Graphs (graph titles) 106
Matching the Missing (graphs) 107
What Does It Mean? (graphs)..................108
Cut and Paste Facts
(sort and classify)..............................109–110
Important Information
(compiling data) 111
Cold Weather Graphs (graphs) 112
Money Graphs (data analysis) 113
A Day at the Fair (data analysis) 114
Three Cheers for the Team!
(representing data) 115

Answer Key.................................116–128

☑ Number and Operations—counting   Name _____

Mix and Match Socks

Count the socks in each basket. Cut out the numbers at the bottom of the page. Glue the right number under the basket. Five answers will not be used.

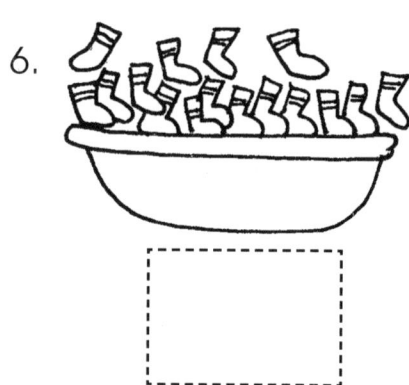

| 22 | 10 | 19 | 7 | 11 | 15 | 4 |
| 12 | 5 | 18 | 8 | 9 | 14 | 16 |

© McGraw-Hill Children's Publishing   IF87125 Standards-Based Math

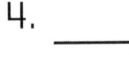 Number and Operations—writing numbers   Name _____

Trace This

Practice tracing each number. Then write each number in the matching space below.

1. 81

2. 102

3. 33

4. 7

5. 94

6. 70

7. 16

8. 55

9. 49

1. _____

2. _____

3. _____

4. _____

5. _____

6. _____

7. _____

8. _____

9. _____

Number and Operations—place value Name _____

Ones and Tens

Look at the numbers. Draw a red circle around each number in the tens place. Draw a blue square around each number in the ones place.

1. 45   2. 14   3. 9   4. 10   5. 66
6. 17   7. 8   8. 41   9. 56   10. 93

Connect the numbers below in order from least to greatest to show an animal who likes bananas.

© McGraw-Hill Children's Publishing   IF87125 Standards-Based Math

Number and Operations—place value   Name _____

A House Full of Place Value

Look at each house. Fill in the numbers on the place value chart above the house. Then color the roofs of houses with numbers between 100 and 200 red. Color roofs of the houses with numbers between 200 and 300 blue. Color the roofs on the houses with numbers between 300 and 400 yellow.

1. hundreds | tens | ones

2. hundreds | tens | ones

3. hundreds | tens | ones

4. hundreds | tens | ones

5. hundreds | tens | ones

6. hundreds | tens | ones

© McGraw-Hill Children's Publishing   IF87125 *Standards-Based Math*

Number and Operations—place value   Name _____

Mystery Numbers

Read the clues and find the numbers.

1. I have a two in the tens place and a seven in the ones.

   Who am I? _____

2. I am between nine and eleven.

   Who am I? _____

3. I am between twenty and thirty. I have a five in the ones place.

   Who am I? _____

4. I am between 100 and 200. I have a zero in the tens place and a three in the ones place.

   Who am I? _____

5. I have a four in the ones, tens, and hundreds places.

   Who am I? _____

6. I have no hundreds or tens. I have an eight in the ones place.

   Who am I? _____

7. I have a three in the hundreds place, a zero in the tens place, and a one in the ones place.

   Who am I? _____

8. I am between ten and fifteen. I have a two in the ones place.

   Who am I? _____

9. I am a number between 40 and 50. I have a six in the ones place.

   Who am I? _____

10. I have a seven in the hundreds place, a nine in the tens place, and an eight in the ones place.

    Who am I? _____

Number and Operations—relative position   Name _____

Finding the "In Between"

In each set, the number in the middle is missing. Write the missing number.

1.
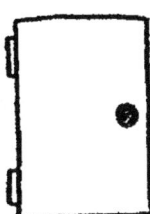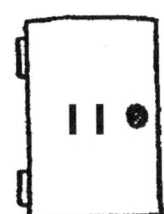

2.

3.

4.

5.

6.

☑ Number and Operations—number words and numerals

Name _____

Teddy Bears

Look at each teddy bear. Do the number and the word match? If they match, color the teddy bear brown. If not, write the correct number and color the bow red.

1.
_____

2.
_____

3.
_____

4.
_____

5.
_____

6.
_____

7.
_____

8.
_____

☑ Number and Operations—fractions   Name _____

Understanding Fractions

Betsy went to the market. Color the fractions of fruits and vegetables that Betsy took home.

1.    Color $\frac{1}{3}$ of the carrots orange.

2.    Color $\frac{1}{4}$ of the apples red.

3.    Color $\frac{1}{3}$ of the potatoes brown.

4.    Color $\frac{1}{2}$ of the oranges orange.

5.    Color $\frac{1}{5}$ of the strawberries red.

6.    Color $\frac{1}{4}$ of the tomatoes red.

7.    Color $\frac{1}{2}$ of the cucumbers green.

8.    Color $\frac{1}{4}$ of the peaches yellow.

☑ Number and Operations—fractions   Name _____

Cupcake Fractions

Look at the cupcakes. Cut out the sentences at the bottom of the page. Glue the sentence that matches each group of cupcakes below each batch.

- -
$\frac{1}{2}$ of the cupcakes have candles.
- -
$\frac{1}{4}$ of the cupcakes do not have hearts.
- -
$\frac{1}{3}$ of the cupcakes have a cherry.
- -

© McGraw-Hill Children's Publishing   IF87125 Standards-Based Math

Number and Operations—fractions   Name _____

The Whole Team

Follow the directions and draw to make each sentence true.

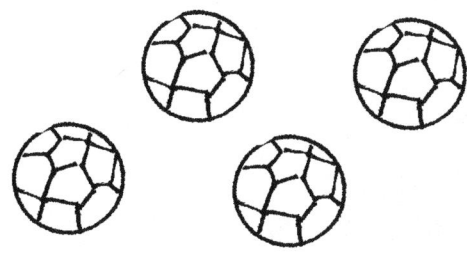

1. Color $\frac{1}{4}$ of the soccer balls black and white.

2. Add a hat to $\frac{1}{2}$ of the tennis players.

3. Add the number 12 to $\frac{1}{4}$ of the football players' shirts.

4. Color $\frac{1}{3}$ of the bats brown.

5. Add black hair to $\frac{1}{2}$ of the hockey players.

6. Add a blue ribbon to $\frac{1}{3}$ of the girls.

Number and Operations—fractions   Name _____

Cut and Paste Fractions

Look at each pie. How much of the pie is colored? Cut out the fractions at the bottom of the page. Glue the right fraction next to each pie.

1.
2.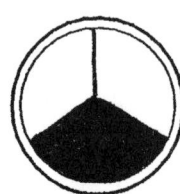
3.
4.
5.
6.
7.
8.
9.
10.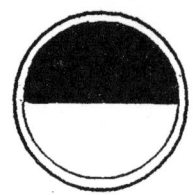

$\frac{1}{3}$	$\frac{1}{4}$	$\frac{1}{2}$	$\frac{1}{5}$	$\frac{1}{10}$
$\frac{1}{2}$	$\frac{1}{4}$	$\frac{1}{5}$	$\frac{1}{2}$	$\frac{1}{3}$

© McGraw-Hill Children's Publishing   IF87125 *Standards-Based Math*

Number and Operations—fractions   Name _____

Show What You Know!

Look at the blocks below. Read the directions. Color a fraction of each set of blocks.

1. 🧊🧊🧊   Color $\frac{1}{3}$ of the blocks green.

2. 🧊🧊🧊🧊🧊   Color $\frac{1}{5}$ of the blocks purple.

3. 🧊🧊🧊🧊🧊🧊🧊🧊🧊🧊   Color $\frac{1}{10}$ of the blocks orange.

4. 🧊🧊   Color $\frac{1}{2}$ of the blocks blue.

5. 🧊🧊🧊🧊   Color $\frac{1}{4}$ of the blocks yellow.

Match each picture to its fraction.

6. ○○○●○○○○○○ $\frac{1}{2}$

7. ■ □ $\frac{1}{4}$

8. □ □ ■ $\frac{1}{3}$

9. ▲ △ △ △ $\frac{1}{5}$

10. △ △ △ △ ▲ $\frac{1}{10}$

☑ Number and Operations—greater than 10   Name _____

Who Is Bigger?

Help the bear get home. Color numbers greater than 10.

Number and Operations—comparison   Name _____

Hungry Hippos

The hippos are very hungry! They want to eat all the big numbers.
Use <, >, or = to make each sentence true. Color the hippo pink where a > is used.

1.  12   2

2.  14   16

3.  3   13

4.  21   12

5.  7   19

6.  11   10

7.  8   18

8.  14   15

9.  31   13

10. 17   17

☑ Number and Operations—addition   Name _____

Compute with Caroline

Help Caroline solve the ten problems. Use the chart to color each key.

1. 3
   + 4

2. 1
   + 9

3. 8
   + 7

4. 4
   + 2

5. 9
   + 8

6. 6
   + 3

7. 7
   + 8

8. 3
   + 9

9. 3
   + 2

10. 8
   + 6

＋   ✕

−   ÷

Color	Sum
green	0–4
blue	5–9
orange	10–12
purple	13–17
brown	18–20

© McGraw-Hill Children's Publishing   20   IF87125 *Standards-Based Math*

☑ Number and Operations—addition   Name _____

Adding It Up

Add. Show the lost hiker the path. Color the stones with sums of 17.

1. 10 + 5 = ___
2. 16 + 8 = ___
3. 4 + 3 = ___
4. 9 + 8 = ___
5. 10 + 7 = ___
6. 8 + 7 = ___
7. 14 + 2 = ___
8. 12 + 5 = ___
9. 3 + 9 = ___
10. 9 + 9 = ___
11. 8 + 9 = ___
12. 4 + 13 = ___
13. 17 + 0 = ___
14. 16 + 1 = ___

✓ Number and Operations—addition   Name _____

Delicious Word Problems

Maya and Lucy are at the bakery. Solve the word problems so they can eat.

1. The baker makes 2 batches of cupcakes. He uses sprinkles on the first batch of 7 cupcakes. He uses frosting on the second batch of 9 cupcakes. What is the total number of cupcakes the baker decorates? _____

2. Sam puts 6 doughnuts in the window. Later he adds 4 doughnuts. How many doughnuts does Sam put in the bakery window in all?
_____

3. Bertha bakes 10 cookies in the morning and 9 cookies in the afternoon. How many cookies does Bertha bake in all? _____

4. A baker puts 6 candles on one cake and 5 candles on a second cake. How many candles are on both cakes? _____

5. The bakery has fruit tarts for sale. They have 9 strawberry tarts and 4 blueberry tarts. How many fruit tarts does the bakery have altogether? _____

☑ Number and Operations—missing addend   Name _____

The Answer Is...

Find the missing numbers. Color by looking at the chart.

Missing Number	Color
0, 2, 4	brown
1, 5, 7	green
3, 6, 9	orange
8, 12	red
10, 11, 13	yellow

1.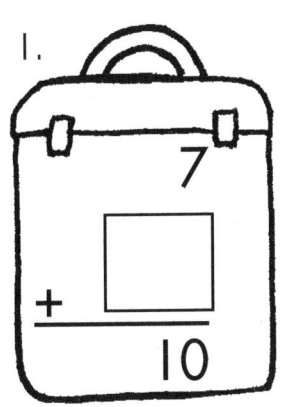
2.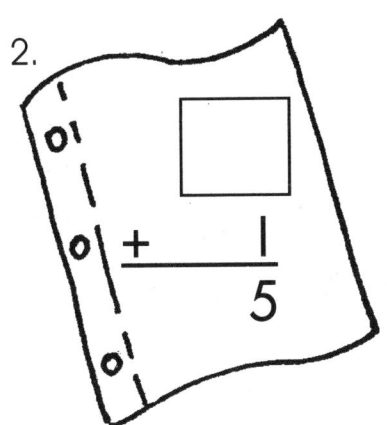
3.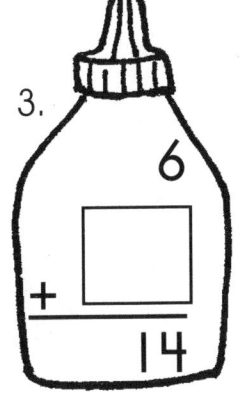
4.
5.
6.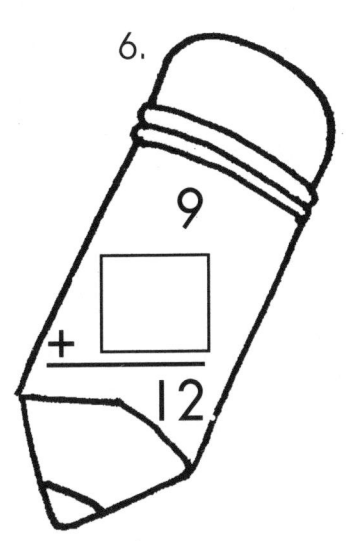
7.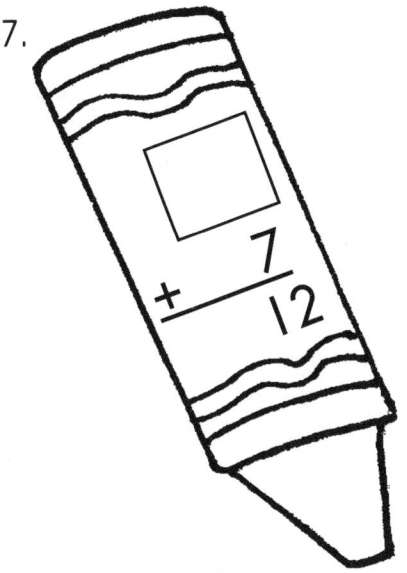
8.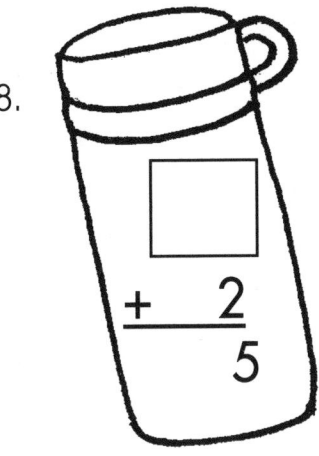

☑ Number and Operations—subtraction   Name _____

Sliding Through Subtraction

Solve each math fact on the slides. Then look at the chart. Color each problem based on the answer.

Difference	Color
0–4	green
5–12	yellow

1. 9 − 3
2. 5 − 4
3. 3 − 2
4. 12 − 5
5. 9 − 6
6. 10 − 2
7. 15 − 9
8. 17 − 6
9. 11 − 4
10. 8 − 7

Number and Operations—subtraction   Name _____

Bouncing Balls

Help Sara stop the bouncing balls. Solve each problem. Look at each ball. Write the number that makes each math fact correct.

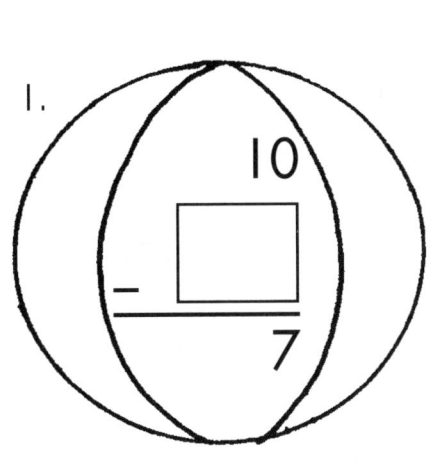
1. 10 − ☐ = 7

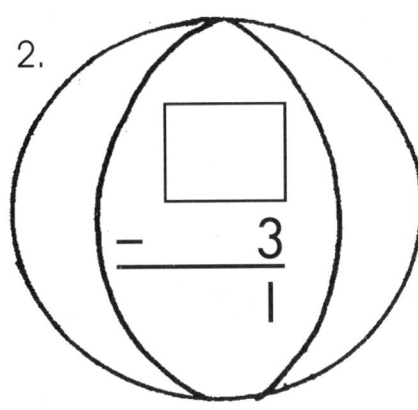
2. ☐ − 3 = 1

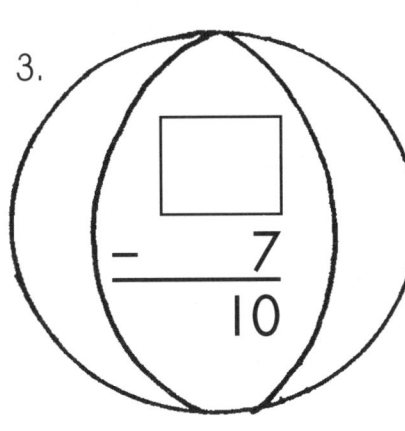
3. ☐ − 7 = 10

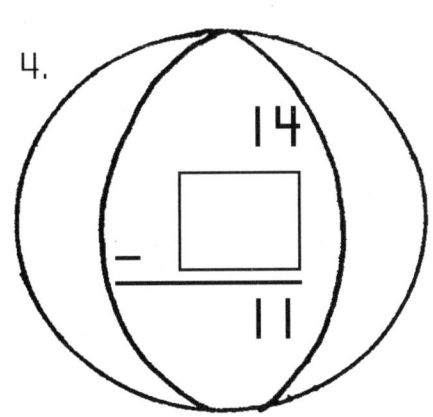
4. 14 − ☐ = 11

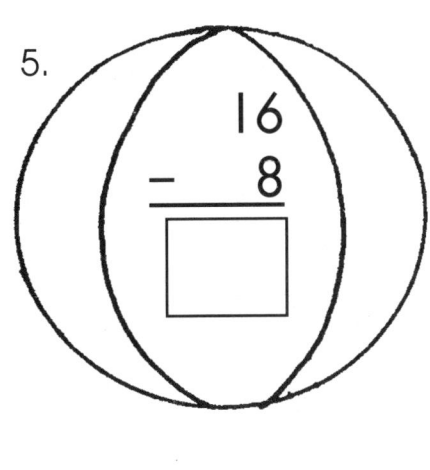
5. 16 − 8 = ☐

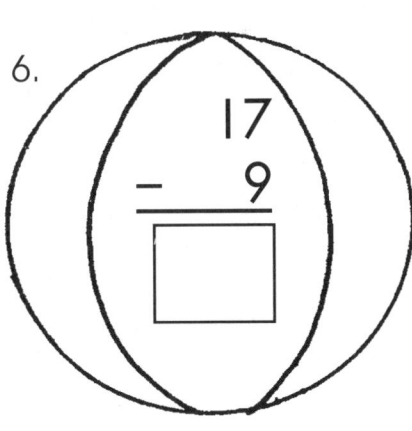
6. 17 − 9 = ☐

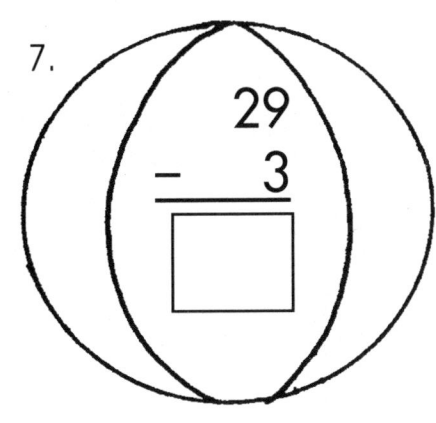
7. 29 − 3 = ☐

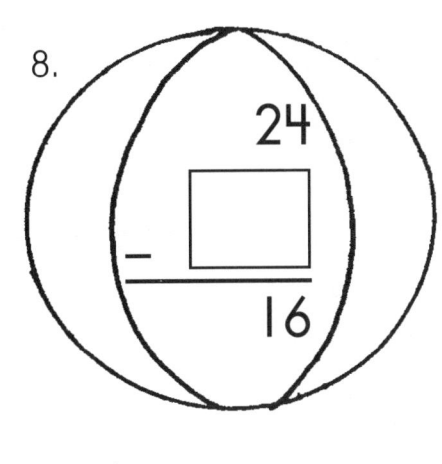
8. 24 − ☐ = 16

☑ Number and Operations—subtraction   Name _____

Subtraction Practice

Subtract. Color spaces with answers greater than 10 pink. Color the rest orange.

1. 22 − 4
2. 21 − 7
3. 18 − 11
4. 4 − 2
5. 10 − 3
6. 14 − 8
7. 25 − 19
8. 18 − 5
9. 25 − 2
10. 24 − 7
11. 20 − 5
12. 15 − 8
13. 7 − 1
14. 14 − 5
15. 6 − 3

© McGraw-Hill Children's Publishing   IF87125 Standards-Based Math

A Bunch of Bubbles

Solve each problem. Color each bubble that has a correct answer.

1. 35 − 14
2. 27 − 17
3. 14 − 7
4. 12 − 8
5. 17 − 3
6. 34 − 3
7. 29 − 18
8. 17 − 3
9. 23 − 12
10. 13 − 12
11. 34 − 22
12. 19 − 4

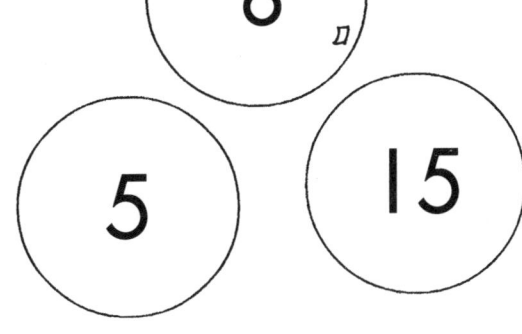

Number and Operations—subtraction   Name _____

Worldly Word Problems

Jack and Janet are going on a trip. They have to solve math problems as they go. Help Jack and Janet. Show your work.

> Sample Problem:
> Jack and Janet packed 4 bags. One bag will not fit. They leave it behind. How many bags do Jack and Janet take?
> They take ___3___ suitcases.   4 – 1 = 3

1. Jack and Janet bought 17 pieces of gum. Before the flight they chewed 6 pieces. How many pieces of gum do Jack and Janet have left?
They have _____ pieces of gum left.

2. Jack and Janet go fishing. Jack catches 8 fish. Janet catches 13 fish. How many more fish does Janet catch?
Janet catches _____ more fish.

3. At the counter, there were 11 people in line. Four people were wearing hats. How many of the people were not wearing hats?
There were _____ people not wearing hats.

4. Janet and Jack buy presents for their friends. Janet spends $9.00. Jack spends $11.00. How much more money does Jack spend?
Jack spent _____ more.

5. Jack and Janet need to take a taxi. There are 16 taxis. Nine taxis are yellow. How many taxis are not yellow?
_____ taxis are not yellow.

© McGraw-Hill Children's Publishing   IF87125 Standards-Based Math

Number and Operations—money   Name _____

Coins

Look at the money. Write the amount of money on the line. Use the chart to color each coin the right color.

1. _____

2. _____

3. _____

4. _____

5. _____

6. _____

Coin	Color
half-dollar	blue
quarter	red
dime	green
nickel	purple
penny	orange

✓ Number and Operations—money   Name _____

Maggie's Wallet

Count the money in each wallet. Match each wallet with the item it can buy on the right.

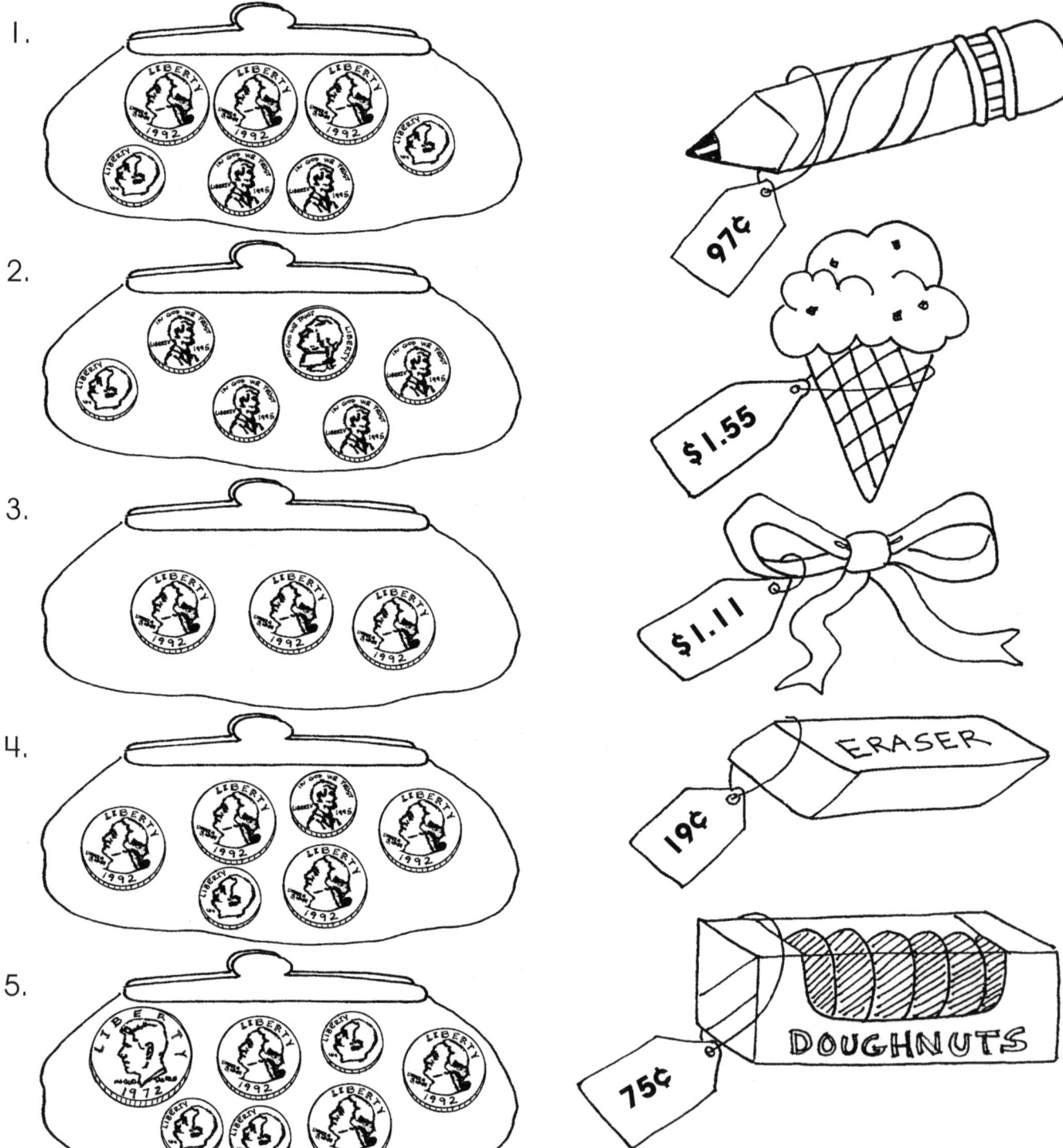

Number and Operations—money   Name _____

A Day at the Mall

Cut out the coins at the bottom of the page. In the space next to each item, glue the coins used to buy it. Use the fewest coins possible.

1.  15¢

2.  91¢

3.  $1.25

4.  49¢

5. ☆ ☆ ☆ ☆ 89¢

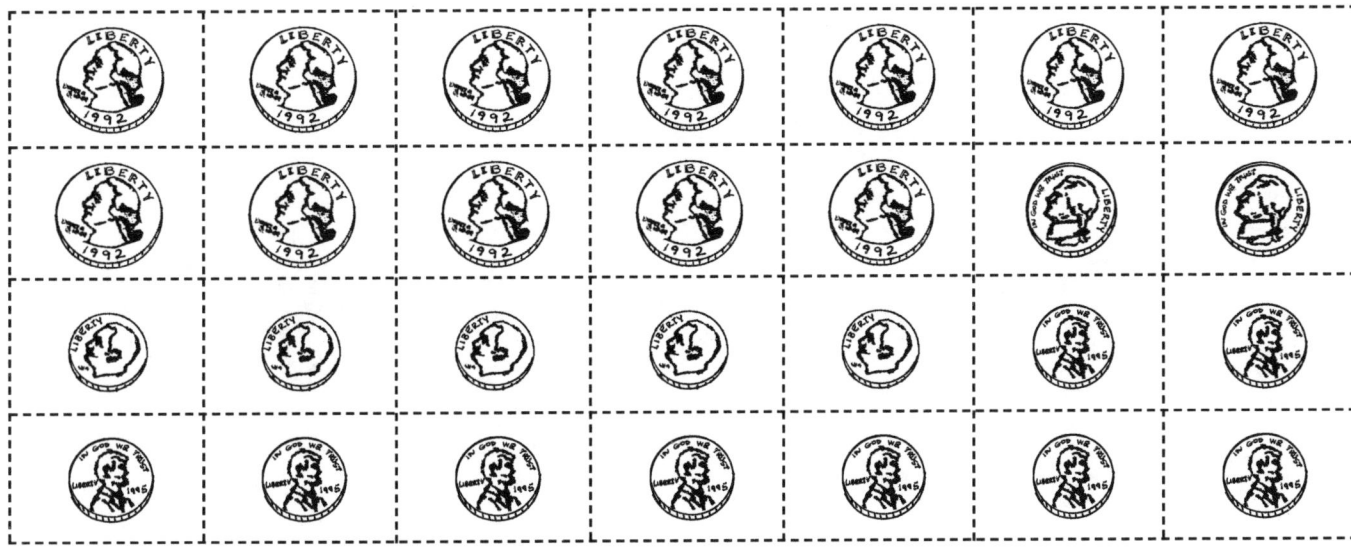

© McGraw-Hill Children's Publishing   IF87125 Standards-Based Math

☑ Number and Operations—money   Name _____

Louie's Loose Change

Write the amount of change Louie has each day.

Monday

Amount _____

Tuesday

Amount _____

Wednesday

Amount _____

Thursday

Amount _____

Friday

Amount _____

Saturday

Amount _____

Sunday

Amount _____

© McGraw-Hill Children's Publishing   IF87125 *Standards-Based Math*

Number and Operations—money   Name _____

Coloring Coins

Write the total of the coins in the box. Color the box according to the chart.

1.

2.

3.

4.

5.

Amount	Color
0¢–35¢	blue
36¢–50¢	orange
51¢–73¢	green
74¢–95¢	yellow
96¢–$1.05	purple
$1.06–$1.25	brown
$1.26–$1.50	red

6.

Number and Operations—addition and subtraction

Name _____

Solve with Words

Solve each problem.

1. Mike the Monkey ate 17 bananas for breakfast, 22 bananas for lunch, and 31 bananas for dinner. How many bananas did he eat in all?

   _____

2. Danny the Dog slept 8 hours Friday night and 10 hours Saturday night. How many more hours did he sleep on Saturday night?

   _____

3. Cassie the Cat ate 19 mice one week and 17 mice the next. How many mice did she eat in two weeks?

   _____

4. There are 201 students at Happy Hills Elementary School. There are 131 girls. How many boys are there?

   _____

5. Kelsey collects baseball cards. She collects 134 cards during first grade and 171 cards during second grade. How many cards does she collect in all?

   _____

6. There are 72 flowers in Abigail's garden. There are 56 sunlight plants. The rest are shade plants. How many of her flowers need to be planted in the shade?

   _____

© McGraw-Hill Children's Publishing   IF87125 Standards-Based Math

✓ Number and Operations—meaning of operations Name _____

Find the Sign

Read each problem. Decide which sign is needed to solve the problem. Write the sign in the box under the problem. Solve the problems.

1. Patsy has 9 dogs and 8 cats. How many dogs and cats does she have altogether?

    ☐

2. Janet spent 5 hours at the mall. She spent 1 hour eating lunch. How much time did Janet spend shopping?

    ☐

3. May has 3 dogs. She has 10 cats. How many pet treats will May hand out?

    ☐

4. Keisha made 10 valentines. Her brother took 3 to give to his friends. How many valentines does Keisha have left?

    ☐

5. Lee studied for 3 hours on Monday, 2 hours on Tuesday, and 3 hours on both Wednesday and on Thursday. How many hours did Lee study in all?

    ☐

6. Pamela has 9 brothers and sisters. She has 5 sisters. How many brothers does she have?

    ☐

7. Jamal practices diving for 3 hours on Monday and on Tuesday, 5 hours on Saturday, and 2 hours on Sunday. How many hours does he practice in all?

    ☐

Number and Operations—meaning of operations   Name _____

Spending and Earning Money

Read each problem. Decide if money is earned or spent. Write an addition sign in the box if money is earned. Write a subtraction sign in the box if money is spent. Write the answer for each problem.

☐ _____ 1. Louise has $7.00. She makes $4.00 for baby-sitting her little sister. How much money does Louise have now?

☐ _____ 2. Cedric has $15.00. He works all day mowing lawns. He gets $12.00 more. How much money does Cedric have now?

☐ _____ 3. Molly has $10.00. She goes out for dinner with friends on Saturday night. She spends $8.00. How much money does Molly have now?

☐ _____ 4. Kamal and Jada save $19.00. They spend $17.00 at the carnival. How much money do they have left?

☐ _____ 5. Jay earns a different amount of money each week based on the chores he does. He earns $7.00 one week, $6.00 the next, and $5.00 the next. How much money does Jay earn in three weeks?

© McGraw-Hill Children's Publishing   IF87125 Standards-Based Math

☑ Number and Operations—2-digit subtraction with regrouping   Name _____

Hats Off to You!

These people lost their hats. Solve the problems on each person. Glue the hat with the right answer on each head.

✓ Number and Operations—2-digit subtraction with regrouping   Name _____

Subtraction Maze

Solve each problem. Color the boxes with differences greater than 39 to solve the maze.

1. 90 − 41
2. 41 − 24
3. 80 − 65
4. 70 − 39

5. 88 − 19
6. 55 − 26
7. 48 − 39

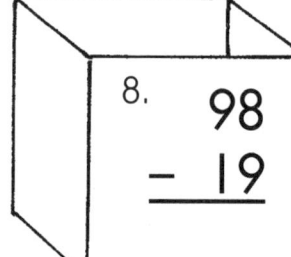

8. 98 − 19
9. 55 − 36
10. 50 − 12
11. 50 − 13

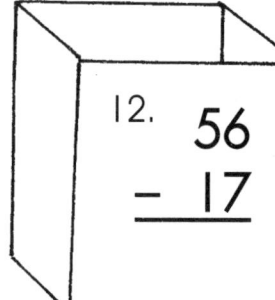

12. 56 − 17
13. 90 − 32
14. 97 − 28

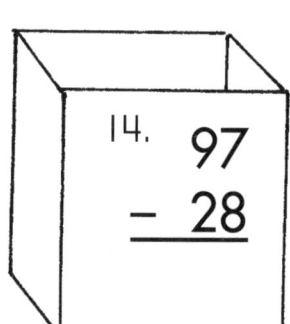

15. 67 − 18

✓ Number and Operations—2-digit subtraction with regrouping Name _____

A Piece of Cake!

Subtract. Color the pieces of cake with answers greater than 25.

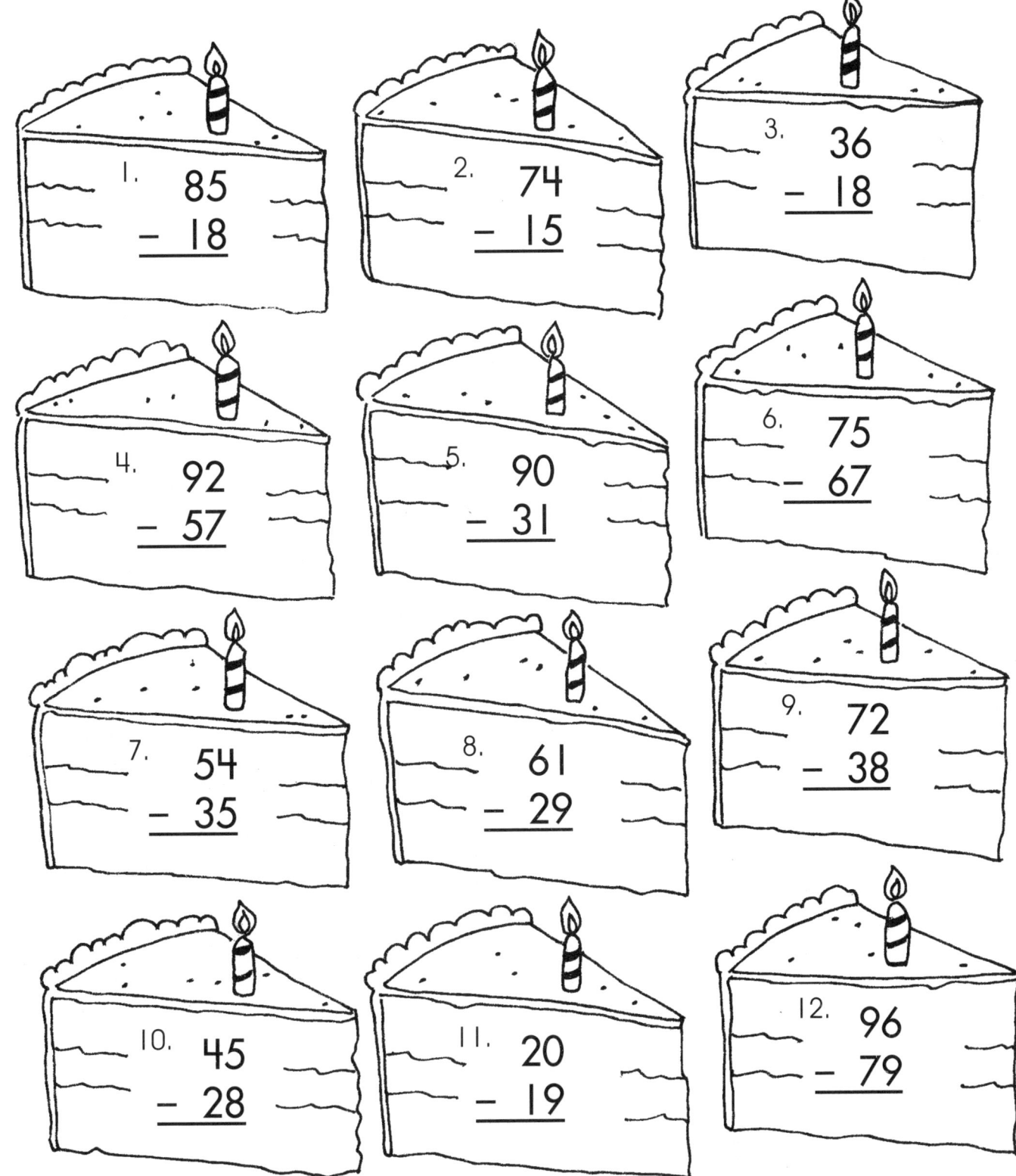

1. 85 − 18
2. 74 − 15
3. 36 − 18
4. 92 − 57
5. 90 − 31
6. 75 − 67
7. 54 − 35
8. 61 − 29
9. 72 − 38
10. 45 − 28
11. 20 − 19
12. 96 − 79

© McGraw-Hill Children's Publishing   IF87125 Standards-Based Math

Number and Operations—2-digit subtraction with regrouping

Name _____

Nutty Solutions

Help the squirrels find the nuts. Solve each problem. Draw a line between each squirrel and the nut that shows the correct answer.

© McGraw-Hill Children's Publishing

IF87125 Standards-Based Math

Cookies, Cookies, Cookies!

Sue is getting ready for the bake sale. Cut out the pictures at the bottom of the page that show how many cookies.

1. There are 3 boxes of cookies. There are 3 cookies in each box.

2. There are 5 boxes of cookies. There are 3 cookies in each box.

3. There are 7 boxes of cookies. There is 1 cookie in each box.

4. There are 2 boxes of cookies. There are 6 cookies in each box.

5. There are 4 boxes of cookies. There are 3 cookies in each box.

6. There are 3 boxes of cookies. There are 5 cookies in each box.

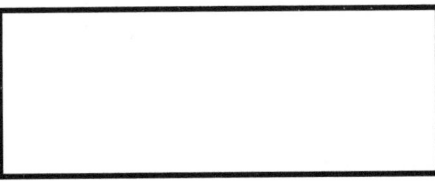

Number and Operations—division/sharing   Name _____

Sharing with Friends

Gabe is sharing. Read each problem carefully. Divide the crayons between the friends. Draw crayons in the boxes to help you. Fill in the correct answer.

1. Gabe divides four crayons between Javier and Andrea. Each student gets _____ crayons.

2. Gabe divides six crayons among three classmates. Each student gets _____ crayons.

3. Gabe divides eight crayons between Jason and Raquel. Each student gets _____ crayons.

4. Gabe divides five crayons among five classmates. Each student gets _____ crayons.

☑ Number and Operations—division/sharing   Name _____

Birthday Party Guests

Ellie invites five friends to her party. She wants to make sure she divides the items equally. Help Ellie write division sentences to make sure.

1. Ellie has 5 pieces of birthday cake.
   Each guest will get _____ piece of cake.

2. Ellie has 15 balloons.
   Each guest will get _____ balloons.

3. Ellie has 10 pieces of gum.
   Each guest will get _____ pieces of gum.

4. Ellie has 5 boxes of juice.
   Each guest will get _____ box of juice.

5. Ellie has 20 straws.
   Each guest will get _____ straws.

Algebra—sort and classify

Name _____

Small to Large

Put the pictures in order from smallest to largest using 1, 2, 3.

1.

    ___   ___   ___

2.

    ___   ___   ___

3.

    ___   ___   ___

4. Draw three stars: one small, one medium, one large.

5. Draw three hearts: one small, one medium, one large.

☑ Algebra—sort and classify   Name _____

Sort It Out

Cut out the pictures below. Sort each one into the correct box.

Objects with 4 sides.

Objects that are cylinders.

Algebra—skip counting   Name _____

Skipping Stones

Lea loves to skip. Help her skip across these stones. Fill in the missing numbers in the patterns.

1. 3, 6, ___, 12, 15, ___, 21

2. ___, 20, ___, 40, 50, ___, 70

3. 2, ___, 6, ___, 10, ___, 14

4. 15, 20, ___, 30, ___, ___, 45

5. 100, ___, 300, ___, 500, ___, 700

6. 12, 18, ___, 30, 36, ___, 48

Algebra—patterning Name _____

Number Patterns

Jung Lee is planting flowers by following number patterns. Find the pattern in each problem. Fill in the missing number.

1. 7, 6, 5, ____, 3
2. 2, 2, 4, 4, 6, ____, 8
3. 0, 4, ____, 12
4. 10, 15, 20, ____
5. 1, 3, ____, 7, 9
6. 6, 12, ____, 24, 30
7. 20, 18, 16, ____, 12
8. 0, 3, 6, 9, ____, 15
9. 8, 16, 24, ____
10. 9, 6, ____, 0

Algebra—patterning Name _____

Square, Square, Circle

Look at the shape patterns. Draw the shape that needs to be added to make each pattern complete.

1. ○ □ △ ○ □ ___

2. □ □ ▭ □ ___

3. ⬡ ⬠ ⬡ ___

4. ___ ○ □ ○ ___

5. ⬠ □ □ ___ □ □

6. ⬡ □ △ ___ □ ___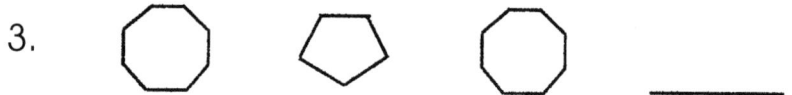

7. ○ ○ ▭ ▭ ___ ○ ▭

8. △ △ ▭ ___ △ △ ▭ ○

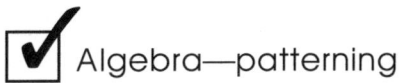 Name _____

Sizable Patterns

Look at each pattern. Connect the missing picture to each pattern. Not all pictures will be used.

1.

2.

3.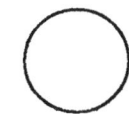

4.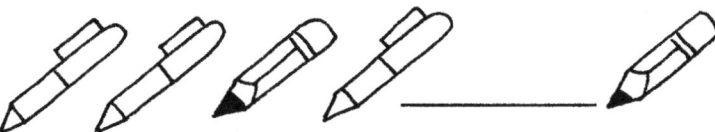

5.

6.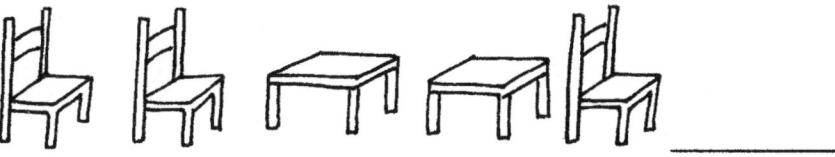

7.

8.

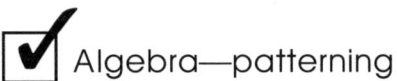 Algebra—patterning

Name _____

Pasting Patterns

Cut out the pictures below. Then look at the patterns. Glue on the picture that makes the pattern complete. You will not use all the pictures.

1.
2.
3.
4.
5.

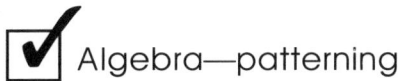 Algebra—patterning   Name _____

Candy Store Patterns

Look at the patterns. Draw in each missing piece of candy.

1.   _____

2.    _____

3.  _____

4. _____

5.   _____

6.  _____

7.     _____

8. _____

Algebra—commutativity   Name _____

It's the Opposite

In an addition problem, the numbers can be switched, but the answer is still the same.

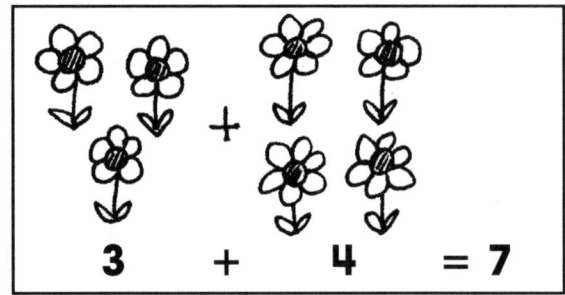

3 + 4 = 7

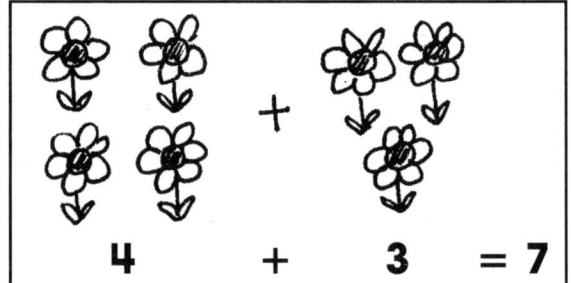

4 + 3 = 7

Complete the problems below.

1.
```
  4        3
+ 3      + 4
```

2.
```
  5        2
+ 2      + 5
```

3.
```
  1        5
+ 5      + 1
```

4.
```
  9        8
+ 8      + 9
```

Rewrite each math fact. Use the pattern above. Solve both.

5.
```
  6
+ 2
```

6.
```
 10
+ 3
```

7.
```
  7
+ 6
```

☑ Algebra—qualitative change   Name _____

Growing Up

Look at how much each child grew. Circle the correct word.

1. Lizzy grew 2 inches.
   Lizzy is _____ .

   a. shorter   b. taller

2. Javier grew 4 inches.
   Javier is _____ .

   a. longer   b. shorter

3. Jenny took off her shoes.
   Jenny is _____ .

   a. taller   b. shorter

4. Michael weighed 80 pounds in third grade. Now Michael is in eighth grade. Michael weighs 125 pounds.
   Michael got _____ .

   a. bigger   b. smaller

5. Janelle's hair was 2 feet long. She cut off 1 foot.
   Janelle's hair is _____ .

   a. longer   b. shorter

☑ Algebra—representing with symbols   Name _____

Solving Symbols

Maggie loves to draw. Pictures help her solve math problems. Solve the problems using the objects shown.

1. 3
   +4   △

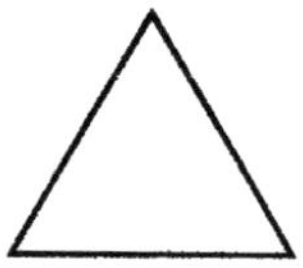

2. 10
   + 2   ◯

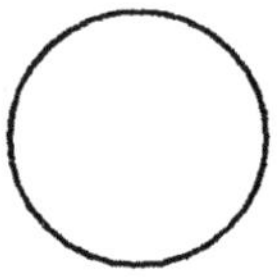

3. 9
   +7   □

4. 8
   +6   ☆

Solve the problems.

5. 6 + △ △ △ △ =

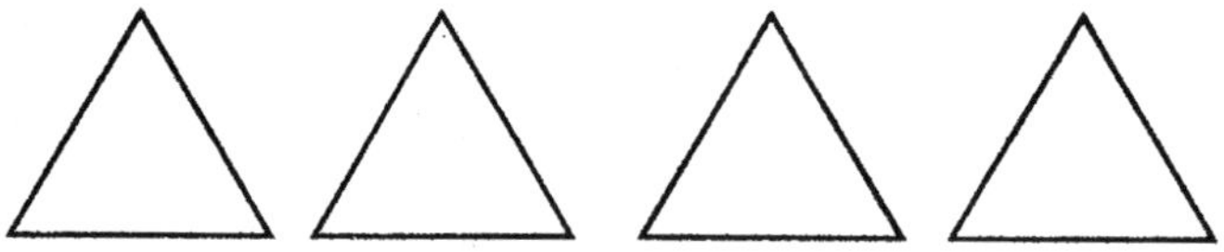

6.  ♡ + 9 =

7.  ☆☆☆☆☆☆☆☆ + 8 =

8. 7 + ☺☺☺☺☺ =

Geometry—shapes   Name _____

Flat Shapes

Trace each shape. Fill in the sentences.

1. 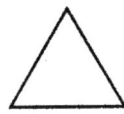　　　　　A triangle has _____ sides.

2. 　　　　　A square has _____ sides.

3. ▭　　　　　A rectangle has _____ sides.

4. ⬠　　　　　A pentagon has _____ sides.

5. ⯃　　　　　An octagon has _____ sides.

Match each shape to its name.

6. ○　　　　　　　　　　　octagon

7. ▭　　　　　　　　　　　pentagon

8. ⬠　　　　　　　　　　　circle

9. 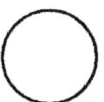　　　　　　　　　　　rectangle

10. ⯃　　　　　　　　　　　hexagon

 Geometry—shapes

Name _____

Flying High

Color the picture. Use the chart to help.

Shape	Color
triangle	black
circle	orange
square	red
rectangle	green
pentagon	yellow
hexagon	purple
octagon	blue

© McGraw-Hill Children's Publishing   IF87125 Standards-Based Math

Geometry—shapes   Name _____

Discovering Details!

Linda and Max are on a shape hunt. They are trying to see how many circles, triangles, squares, rectangles, pentagons, hexagons, and octagons they can find. Help Linda and Max find the shapes. Fill in your answers below.

How many of each shape did you find?

1. I found _____ circles.
2. I found _____ triangles.
3. I found _____ squares.
4. I found _____ rectangles.
5. I found _____ pentagons.
6. I found _____ hexagon.
7. I found _____ octagon.

© McGraw-Hill Children's Publishing   IF87125 Standards-Based Math

Geometry—3-D shapes   Name _____

Three-Dimensional Shapes

These are three-dimensional shapes. Some are curved, like the sphere, and some have faces, like the cube. Trace each shape.

1.    2.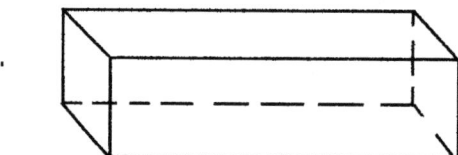

3. 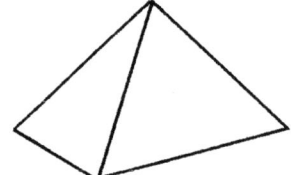   4.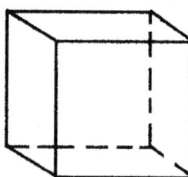

5. 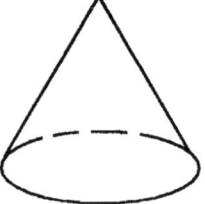   6.

Match each shape to its name.

7.    cylinder

8.   cube

9.   prism

10.   pyramid

11.   cone

12.   sphere

© McGraw-Hill Children's Publishing   IF87125 Standards-Based Math

 Geometry—shapes

Name _____

In the Air

How many of each shape can you find in the picture? Color each shape and record your answers.

1. I found _____ circles.
2. I found _____ triangles.
3. I found _____ squares.
4. I found _____ rectangles.
5. I found _____ pentagon.
6. I found _____ hexagons.
7. I found _____ octagon.
8. I found _____ spheres.
9. I found _____ prisms.
10. I found _____ pyramid.
11. I found _____ cube.
12. I found _____ cones.
13. I found _____ cylinder.

 Geometry—shapes

Name _____

Shapes Take Off

Cut out the shapes shown below. Glue each shape next to its name.

[ ] 1. square	[ ] 2. circle
[ ] 3. sphere	[ ] 4. rectangle
[ ] 5. cube	[ ] 6. octagon
[ ] 7. cone	[ ] 8. pentagon
[ ] 9. cylinder	[ ] 10. hexagon

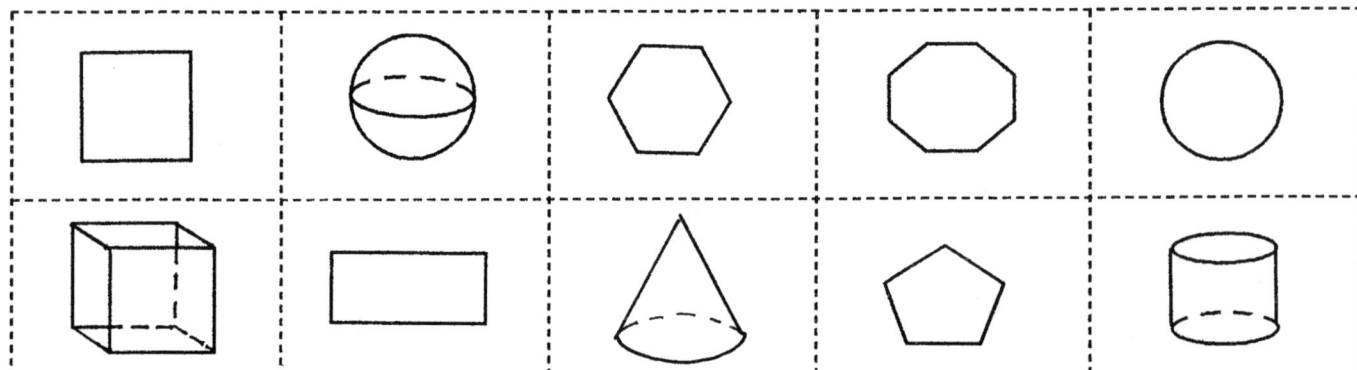

© McGraw-Hill Children's Publishing — IF87125 Standards-Based Math

☑ Geometry—shapes   Name _____

Who Am I?

Use the clues to guess each shape. Draw a line to the correct shape. Then write the name of the shape on the line.

1. I am a three-dimensional shape. I have curved sides. I have no faces. Who am I?
   _____

2. I am a flat shape. I have four sides. My sides are not all equal. Who am I?
   _____

3. I am a three-dimensional shape. I have six faces. Each face is a square. Who am I? _____

4. I am a flat shape. I have no sides. Who am I?
   _____

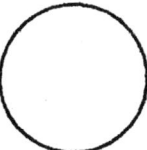

5. I am a three-dimensional shape. Two faces are circles. Who am I?
   _____

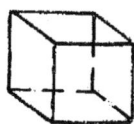

cube   cylinder   rectangle   circle   sphere

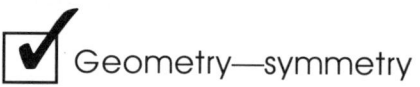 Geometry—symmetry   Name _____

Symmetry

When an object has symmetry, one side looks the same as the other. Look at the ten objects below. Circle **Symmetrical** or **Not Symmetrical** under each object. The first one is done for you.

1.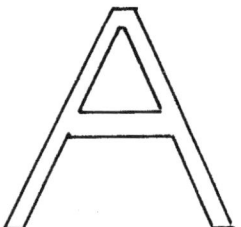

   (Symmetrical)   Not Symmetrical

2.

   Symmetrical   Not Symmetrical

3.

   Symmetrical   Not Symmetrical

4.

   Symmetrical   Not Symmetrical

5.

   Symmetrical   Not Symmetrical

6.

   Symmetrical   Not Symmetrical

7.

   Symmetrical   Not Symmetrical

8.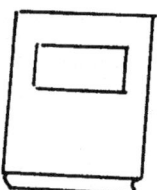

   Symmetrical   Not Symmetrical

Geometry—symmetry   Name _____

Find My Match

Cut out the pictures at the bottom of the page. Look at the pictures below. Glue the other half of the picture to make the object symmetrical.

1.

2.

3.

4.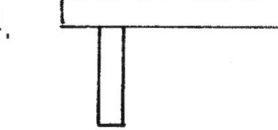

5.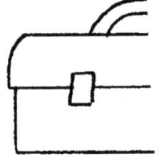

6.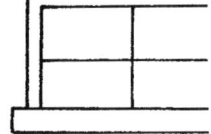

7.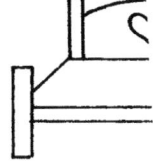

8.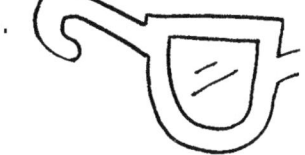

© McGraw-Hill Children's Publishing   79   IF87125 *Standards-Based Math*

Geometry—symmetry

Name _____

Simply Symmetrical

Look at the picture. Find 11 objects with symmetry. Color them red.

Geometry—symmetry Name _____

Mirror, Mirror, on the Wall

Look at each picture. Draw the other half of each object so it is symmetrical.

1.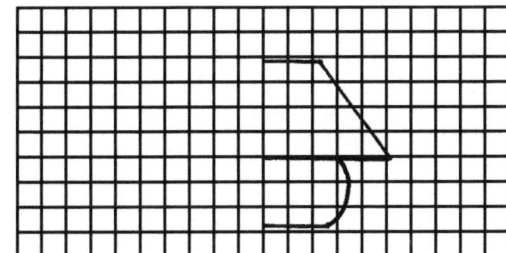

2.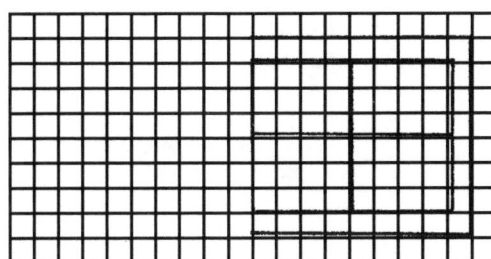

3.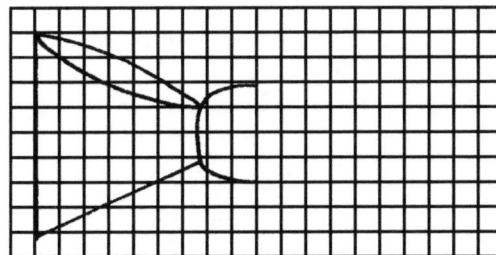

4.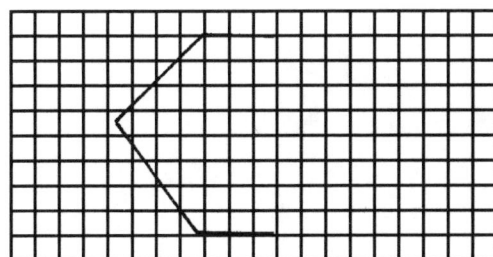

5.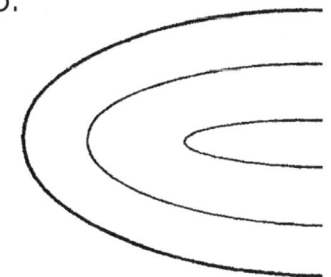

6.

7.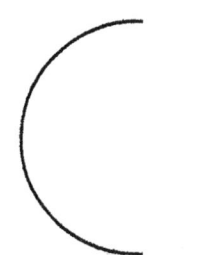

8.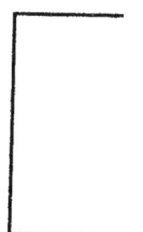

Geometry—symmetry Name _____

Lines of Symmetry

A line of symmetry shows where an object should be folded so that it has symmetry. Circle the objects below that show a correct line of symmetry.

1.
2.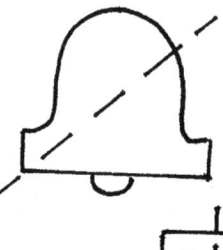
3.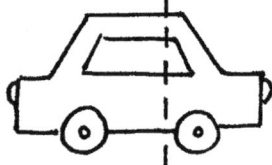
4.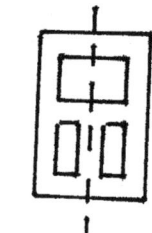
5.
6.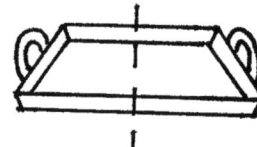
7.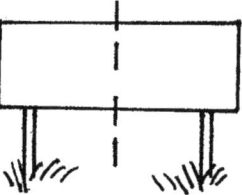
8.

Color all symmetrical objects purple. Color all the objects that are not symmetrical green.

9.
10.
11.
12.
13.

© McGraw-Hill Children's Publishing   IF87125 Standards-Based Math

 Geometry—flips

Name _____

Double Take

Match each picture on the left to a flipped picture on the right.

1.

2.

3.

4.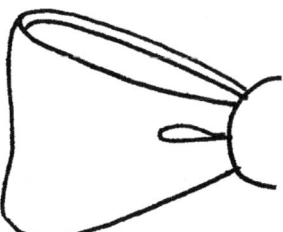

5.

6.

© McGraw-Hill Children's Publishing
IF87125 Standards-Based Math

Geometry—shapes　　　　　　　　　　Name _____

Shapes Around Us

Look at the shapes. Write the name of each shape on the line. Then follow the directions below. Use another piece of paper.

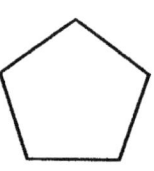

1. _____

2. _____

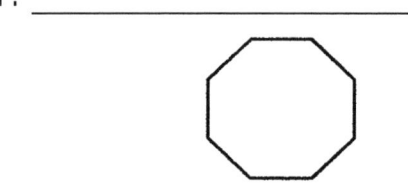

3. _____

4. _____

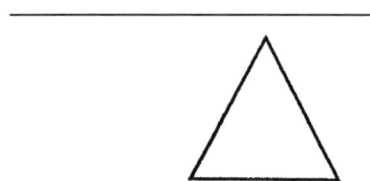

5. _____

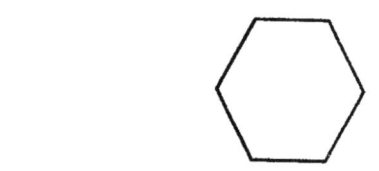

6. _____

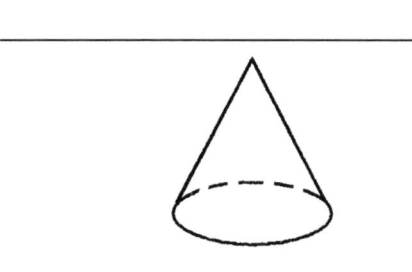

7. _____

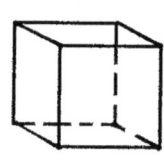

8. _____

9. Draw an outdoor scene using an octagon and three cones.
10. Draw a school scene using a pentagon and two rectangles.
11. Draw a picture using two rectangles and a hexagon.
12. Draw a picture using a triangle, pentagon, and an octagon.
13. Draw a picture using a cone, an octagon, and a pentagon.

© McGraw-Hill Children's Publishing　　　　IF87125 Standards-Based Math

Geometry—symmetry   Name _____

Symmetry Review

Complete and color each picture so that each shape is symmetrical.

1.

2.

3.

4.

5.

6.

7.

8.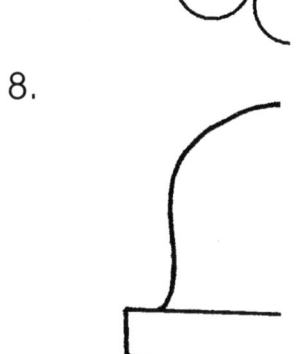

© McGraw-Hill Children's Publishing   85   IF87125 Standards-Based Math

✓ Measurement—nonstandard measurement   Name _____

Counting Caterpillars

How long is each caterpillar? Count sections. Write your answer on the line. Use the chart to color each caterpillar the right color.

Length	Color
1–2	green
3–4	yellow
5–6	orange
7–8	red

1.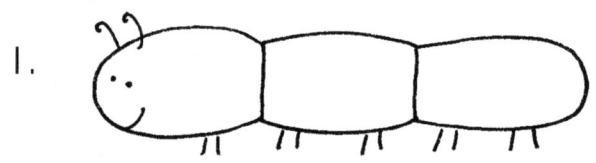

   _____

2.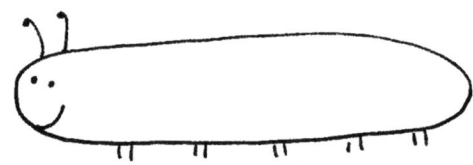

   _____

3.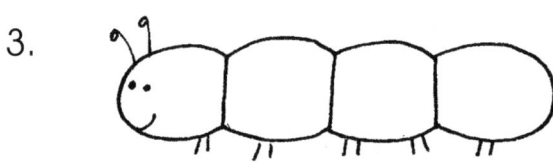

   _____

4.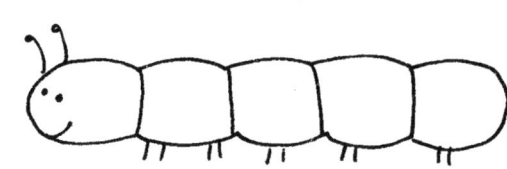

   _____

5.

   _____

6.

   _____

7.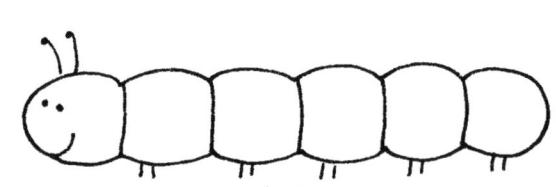

   _____

8.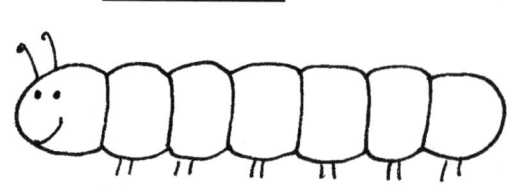

   _____

© McGraw-Hill Children's Publishing   IF87125 Standards-Based Math

Measurement—nonstandard measurement   Name _____

Paper Clip It

Use a 1¼-inch (3.2 mm) paper clip to measure each object below.

1.

_____ paper clip(s)

2.

_____ paper clip(s)

3.

_____ paper clip(s)

4.
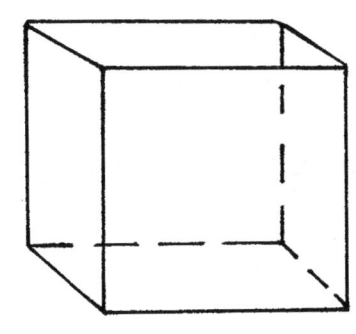

_____ paper clip(s)

5.
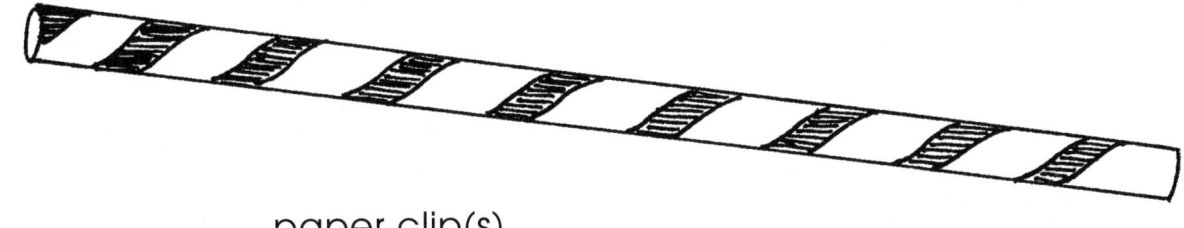

_____ paper clip(s)

6.

_____ paper clip(s)

 Measurement—inches

Name _____

Looking for Length

Use a ruler to measure each object in inches.

1. _____ inches

2. _____ inches

3. _____ inches

4. _____ inches

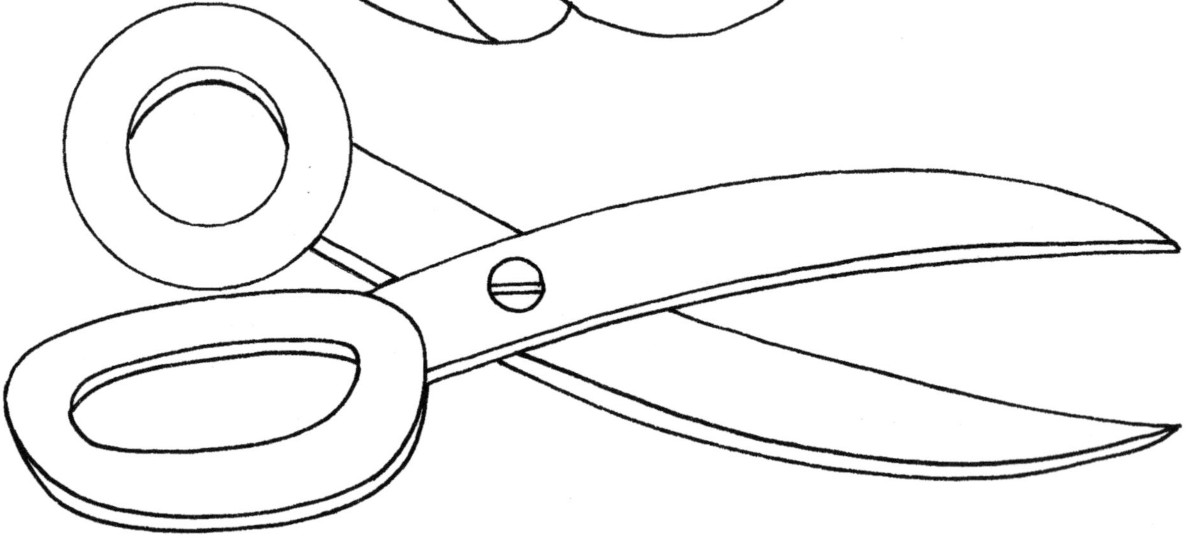

5. _____ inches

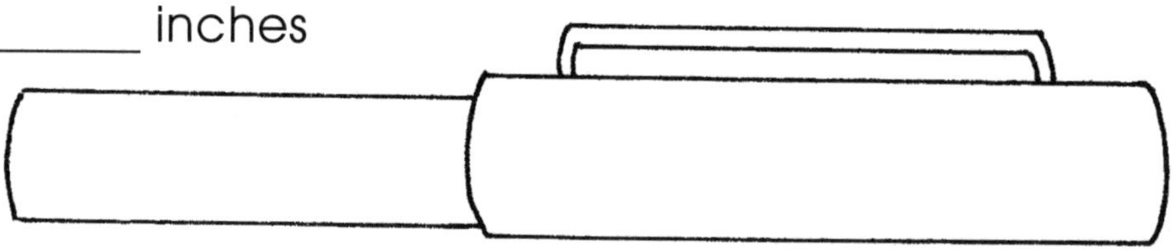

 Measurement—inches

Name _____

Taking the Measurement

Use your ruler to measure the height of each cowboy hat in inches. Then color each hat according to the chart.

Height	Color
0–1	brown
2–3	green

1. _____ inches

3. _____ inches

2. _____ inches

5. _____ inches

4. _____ inches

© McGraw-Hill Children's Publishing

IF87125 *Standards-Based Math*

Measurement—centimeters/meters   Name _____

Traveling with Squirrel

Help Squirrel get his nuts home. Read the map and answer the questions.

1. How far did Squirrel go before he saw the mean dog? _____ m

2. How far did Squirrel have to go before he got to the nice lady? _____ m

3. Squirrel rests at the nice lady's house. How much farther does he have to go? _____ m

4. How far must Squirrel leap to get from the ground to his tree? _____ m

5. How far does Squirel travel in all? _____ m

Measurement—centimeters   Name _____

Munching

Write the length of each piece of licorice. Subtract to show how much was eaten.

1.

☐ cm
− ☐ cm
_____

2.

☐ cm
− ☐ cm
_____

3.

☐ cm
− ☐ cm
_____

4.

☐ cm
− ☐ cm
_____

5.

☐ cm
− ☐ cm
_____

6.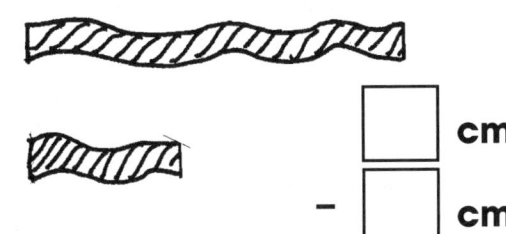

☐ cm
− ☐ cm
_____

 Measurement—pounds Name _____

What Is the Wacky Weight?

Write the weight of each object next to the scale. Then look at the picture of Lyle's closet. Color the objects that weigh 5 pounds or more blue. Color all the objects that weigh less than 5 pounds red.

1.  _____
2. _____ 
3. _____ 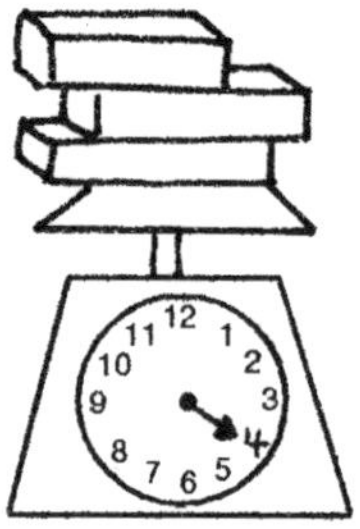
4.  _____
5.  _____
6. _____ 
7. _____ 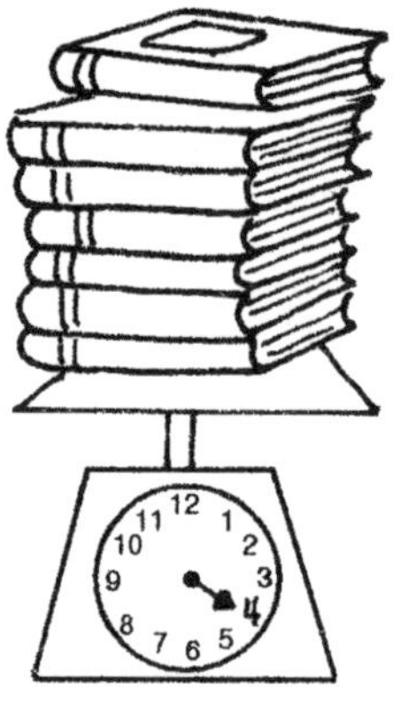
8. _____

© McGraw-Hill Children's Publishing

Measurement—kilograms   Name _____

Basket Weight

Write the weight of each basket. Remember, these scales weigh in kilograms.

1.

   _____ kg

2.

   _____ kg

3.

   _____ kg

4.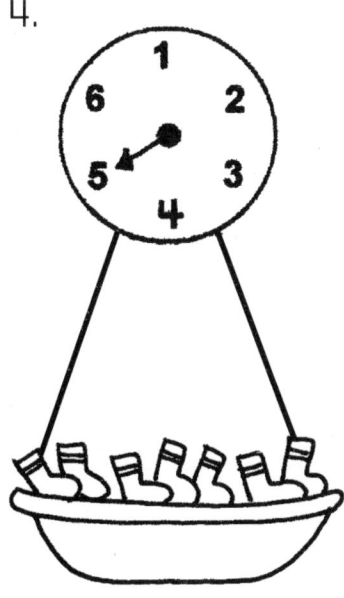

   _____ kg

5.

   _____ kg

6.

   _____ kg

All About Area

Look at each box. Find the area. The first one is done for you.

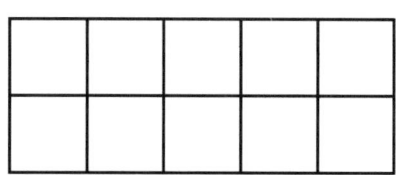

1. __10 square units__

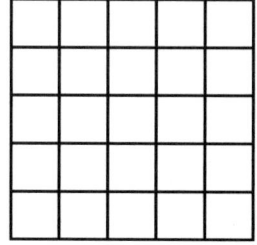

2. _____

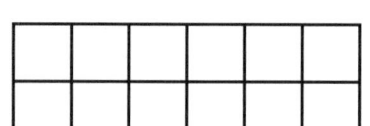

3. _____

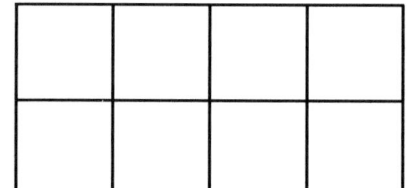

4. _____

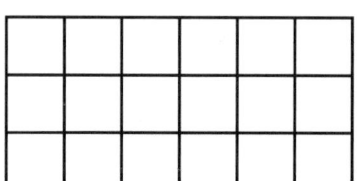

5. _____

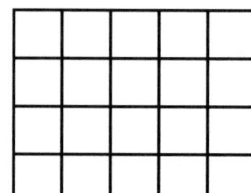

6. _____

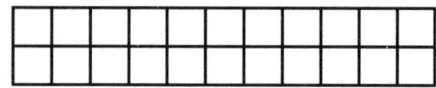

7. _____

8. _____

Measurement—length, weight, area Name _____

Measurement Review

Review what you have learned about measurement. Write the length, weight, or area.

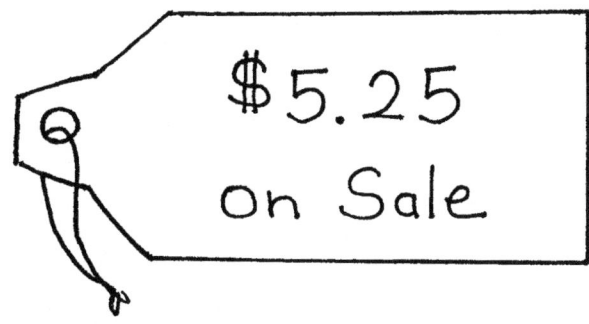
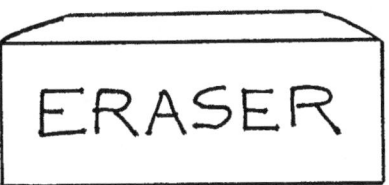

1. _____ inches 2. _____ inches

3. _____ inches

4. _____ pounds 5. _____ pounds

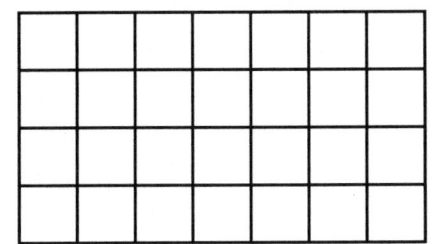
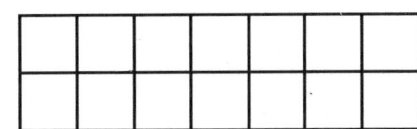

6. _____ sq. units 7. _____ sq. units

© McGraw-Hill Children's Publishing   IF87125 Standards-Based Math

Measurement—cups, pints, quarts   Name _____

At the Race

Lucy's running club poured water at the race. Use the drawings to help you answer the questions.

1 cup

2 cups
1 pint

2 pints
1 quart

1. Lucy's club poured 40 cups of water. They gave away 36 cups. How much was left?

    1 cup   1 pint   1 quart

2. One runner drank 1 cup during the race. He came back for 3 more cups. How much did he drink in all?

    1 pint   2 pints   1 quart

3. In one hour, the workers gave away 2 quarts of water. How many cups did they give away?

    2 cups   4 cups   8 cups

4. Keisha spilled 8 cups of water. How much did she spill?

    4 pints   4 quarts   1 quart

Measurement—time   Name _____

Telling Time

Look at each clock. Write the letter of each clock above the matching time at the bottom of the page to answer the riddle.

Is your refrigerator running? Well...

1.
2.
3.
4.
5.
6.
7.
8.
9.
10.
11.
12.

___ ___ ___   ___ ___ ___   ___ ___ ___   ___ ___
4:00 2:00 3:00 6:00 12:30 9:30 9:30 12:30 5:00 1:30 2:00

                                                      ___!
___ ___ ___ ___ ___   ___ ___
5:30 7:00 9:30 5:30 8:30 3:30 9:30

© McGraw-Hill Children's Publishing   IF87125 Standards-Based Math

Measurement—time

Name _____

The Zookeeper's Watch

The zookeeper is setting watches around the zoo. The animals must be fed at certain times. Read the sentences below. Fill in the time on each watch.

1. He feeds the bears at 10:15 a.m.

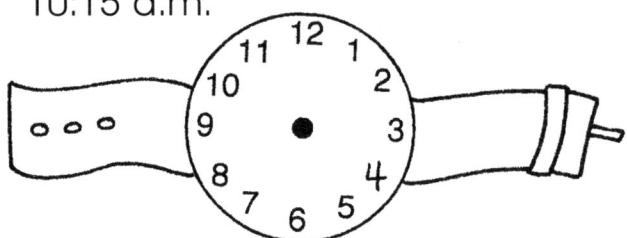

2. The lions must be fed at 9:30 a.m.

3. The monkeys want bananas at 12:30 p.m.

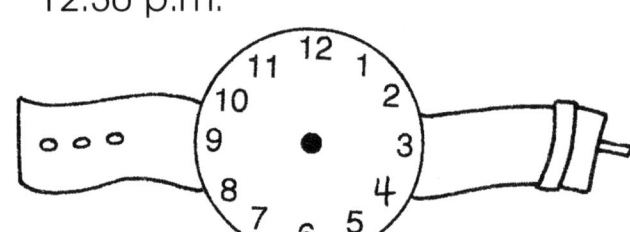

4. He eats his own lunch at 11:30 a.m.

5. He feeds the tigers at 10:00 a.m.

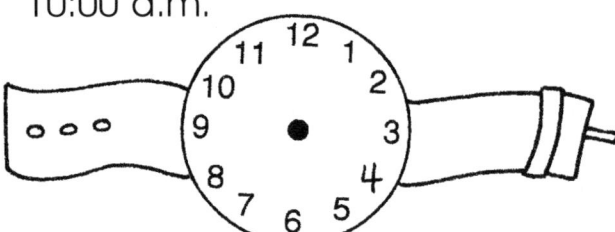

6. The camels are fed at 5:30 p.m.

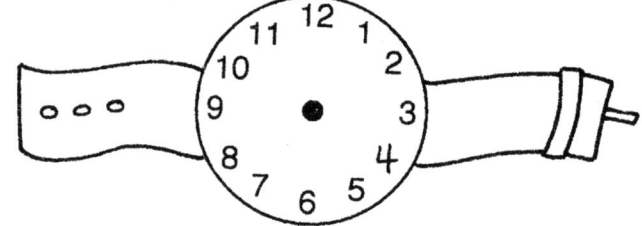

7. He stops by the bird area at 11:00 a.m.

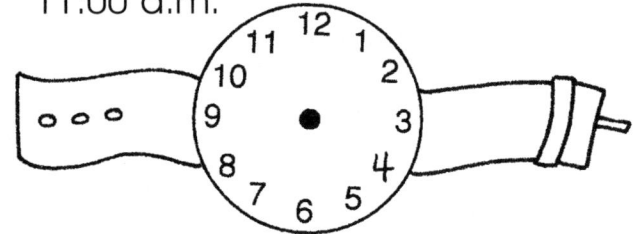

8. He visits the alligators at 4:00 p.m.

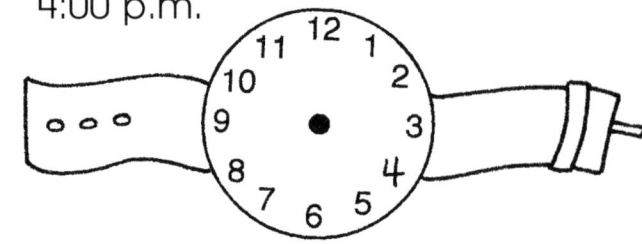

© McGraw-Hill Children's Publishing   IF87125 Standards-Based Math

Measurement—time

Name _____

Don't Be Late!

Harry is busy. Look at his schedule. Look at the six clocks. Write the correct time on the first line under the clock. Then write the activity on the second line.

10:30	AM	Shop
11:00	AM	Soccer
1:00	PM	Read
3:30	PM	Party
5:00	PM	Football
8:00	PM	Bath

1. _____
   _____

2. _____
   _____

3. _____
   _____

4. _____
   _____

5. _____
   _____

6. _____
   _____

© McGraw-Hill Children's Publishing   IF87125 *Standards-Based Math*

☑︎ Measurement—time Name _____

Watching the Time

Draw the hands on each clock to show the times below.

1. 1:15   2. 8:55   3. 2:45

 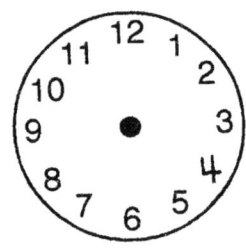 

4. 6:30   5. 9:10   6. 5:15

7. 7:35   8. 10:20   9. 11:00

© McGraw-Hill Children's Publishing   IF87125 Standards-Based Math

Measurement—time   Name _____

Time Puzzlers

Help Francine. Read the six clues below. Write the name of the place or activity under each clock.

1.

    _____

2.

    _____

3.

    _____

4.

    _____

5.

    _____

6.

    _____

Brunch with Brenda
Jogging
Storytime
Tennis
Grocery shop
Bedtime

1. Francine always jogs at half past 8.
2. She is meeting Brenda at quarter after 10.
3. Storytime at the library is one hour before noon.
4. Francine's tennis lesson is at half past 3.
5. She plans to go to the grocery store at a quarter before 5.
6. Francine goes to bed 3 hours before midnight.

Measurement—minutes, hours, days, weeks, months

Name _____

Vacation Time!

Read the story problems and solve. Use the calendars for help.

June

S	M	T	W	Th	F	Sa
					1	2
3	4	5	6	7	8	9
10	11	12	13	14	15	16
17	18	19	20	21	22	23
24	25	26	27	28	29	30

July

S	M	T	W	Th	F	Sa
1	2	3	4	5	6	7
8	9	10	11	12	13	14
15	16	17	18	19	20	21
22	23	24	25	26	27	28
29	30	31				

1. Mary went to camp for 2 weeks. How many days was she gone?

   _____

2. Avishai missed 61 days of school. About how many months did she miss?

   _____

3. Roberto's tennis camp met from June 1 to July 31? How many months was Roberto at camp?

   _____

4. During vacation, Abby read for 72 hours. If she had read all that time at once, for how many days would she have read?

   _____

5. Chin took 2 weeks of vacation in May, 2 weeks in July, 2 weeks in November, and 2 weeks in December. How many months in all did Chin vacation?

   _____

6. Yaron and Sara waited for the plane for 15 minutes. For what part of an hour did they wait?

   _____

© McGraw-Hill Children's Publishing   IF87125 Standards-Based Math

✓ Data Analysis and Probability—tally chart   Name _____

Taking Notes

Jesse wants to know what soda his classmates like best. Use the sentences and tally chart to help him find out. Then decorate the two soda cans.

1. Patricia, Laura, and Patrick vote for lemon lime.
2. Nine children pick cola as their favorite flavor.
3. Gertrude, Gwen, Gabe, and Gweneth like grape soda best.
4. Fourteen students pick orange as their favorite flavor.
5. Twelve students agree with Patricia, Laura, and Patrick.
6. Six students agree with Gertrude, Gwen, Gabe, and Gweneth.
7. Fifteen students pick root beer as their favorite flavor.

root beer	orange	grape	cola	lemon lime

© McGraw-Hill Children's Publishing   IF87125 Standards-Based Math

☑ Data Analysis and Probability—bar graph   Name _____

On the Chart

Use page 103 to help fill in the bar graph below. Use the colors listed below each flavor to make the bar. Use the information to finish the sentences.

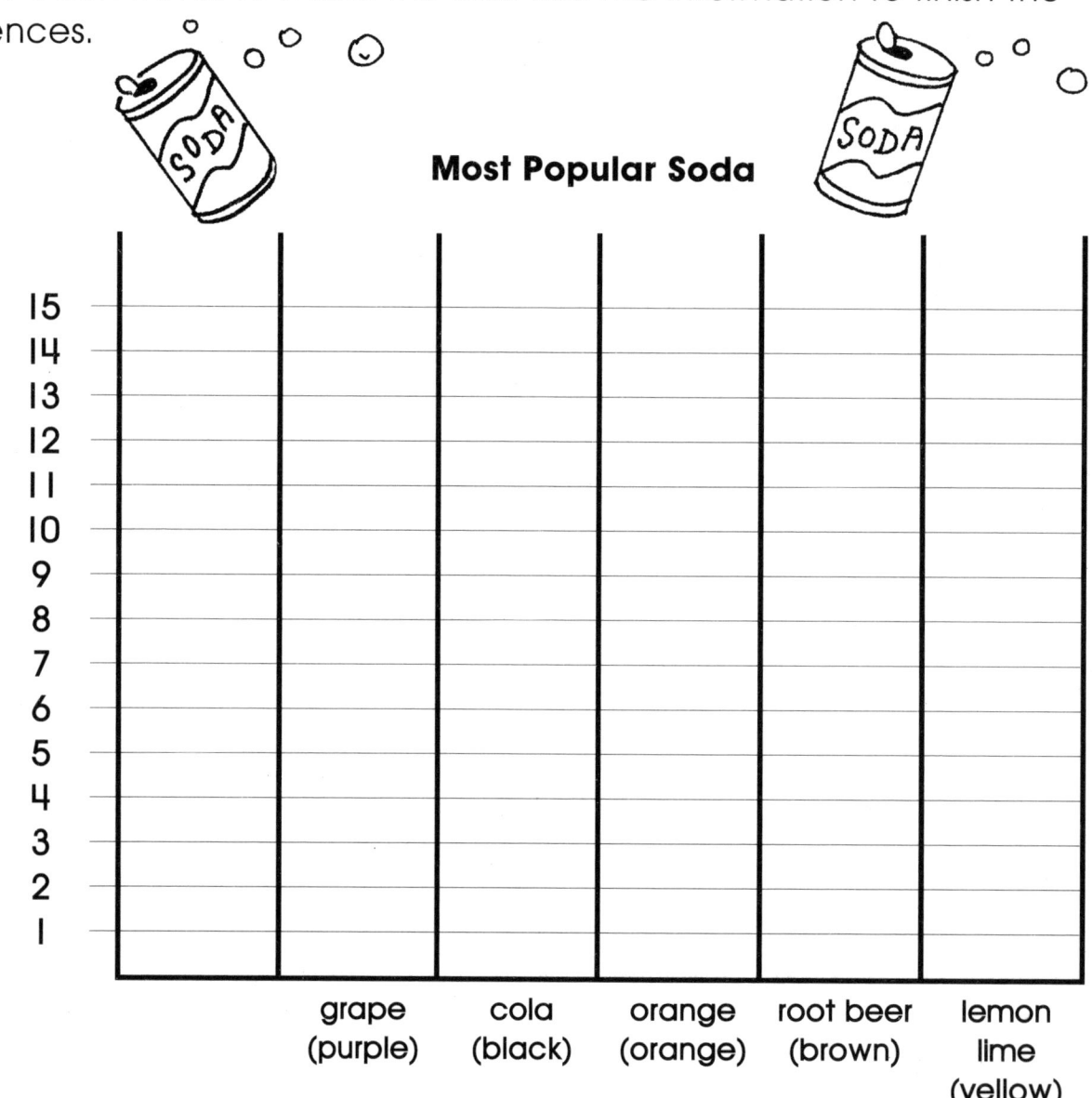

Most Popular Soda

15
14
13
12
11
10
9
8
7
6
5
4
3
2
1

grape (purple)   cola (black)   orange (orange)   root beer (brown)   lemon lime (yellow)

1. The most popular sodas were _____ and _____.
2. Fourteen students voted for _____.
3. Nine students voted for _____.

© McGraw-Hill Children's Publishing   104   IF87125 *Standards-Based Math*

☑ Data Analysis and Probability—pictograph   Name _____

Pizza Pictograph

Mrs. Harris asks her students to vote for their favorite type of pizza. She uses a graph to display the information. Write the number of students who voted for each type of pizza at the end of the row. Then answer the five questions below. Remember, if 1 pizza = 4 students, then $\frac{1}{2}$ a pizza = 2 students.

Pizza Topping Preference

Pepperoni	🍕🍕🍕🍕	____
Hamburger	🍕🍕🍕	____
Sausage	🍕🍕🌗	____
Vegetable	🍕🍕🍕🌗	____

 = 4 students

1. How many students like pepperoni best?

   _____

2. Altogether, how many students voted for vegetable and sausage?

   _____

3. How many more students voted for vegetable than sausage?

   _____

4. What was the total number of students that voted for hamburger, sausage, and vegetable?

   _____

5. What was the least popular type of pizza?

   _____

© McGraw-Hill Children's Publishing   IF87125 Standards-Based Math

Data Analysis and Probability—graph titles   Name _____

Guessable Graphs

The graphs below are missing their titles. Look carefully at each graph. Match it with the title that makes sense. Two of the titles will not be used.

1.

purple	🖍🖍🖍🖍
green	🖍🖍
red	🖍🖍🖍🖍🖍
orange	🖍🖍🖍

🖍 = 2 students

Favorite Colors

2.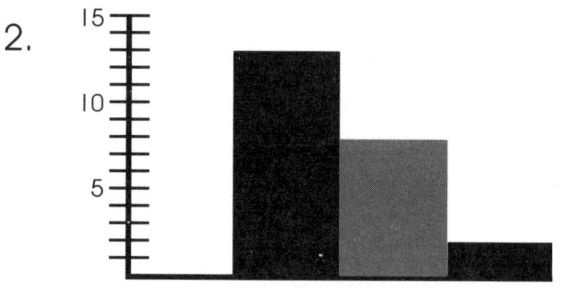

Weekly Telephone Calls

Most Watched TV Show

3.

Monday	☎
Tuesday	☎ ☎
Wednesday	☎ ☎
Thursday	
Friday	☎ ☎

☎ = 4 calls

Number of Books Read by Second Graders

Favorite Ice Cream Flavors

4.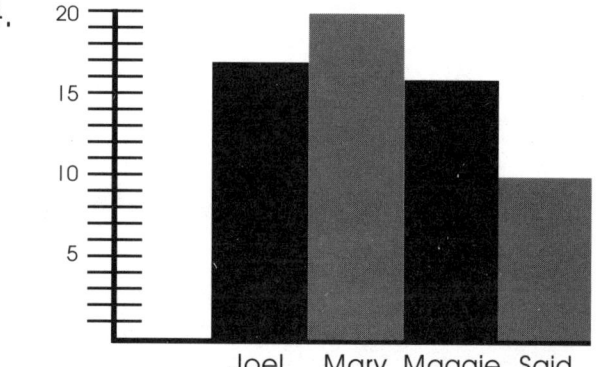

Inches of Rainfall for 1999–2000

© McGraw-Hill Children's Publishing   IF87125 Standards-Based Math

Data Analysis and Probability—graphs   Name _____

Matching the Missing

Cut out the ice cream cones. Glue them on the pictograph so the correct information is shown. You may need to cut an ice cream cone in half. Then answer the questions.

Favorite Ice Cream Flavors

Chocolate (12)	
Vanilla (9)	
Strawberry (18)	
Cookie Dough (15)	

🍦 = 6 students

1. How many more students like chocolate than vanilla? _____
2. How many students voted for vanilla and strawberry? _____
3. How many students voted for their favorite ice cream flavor? _____

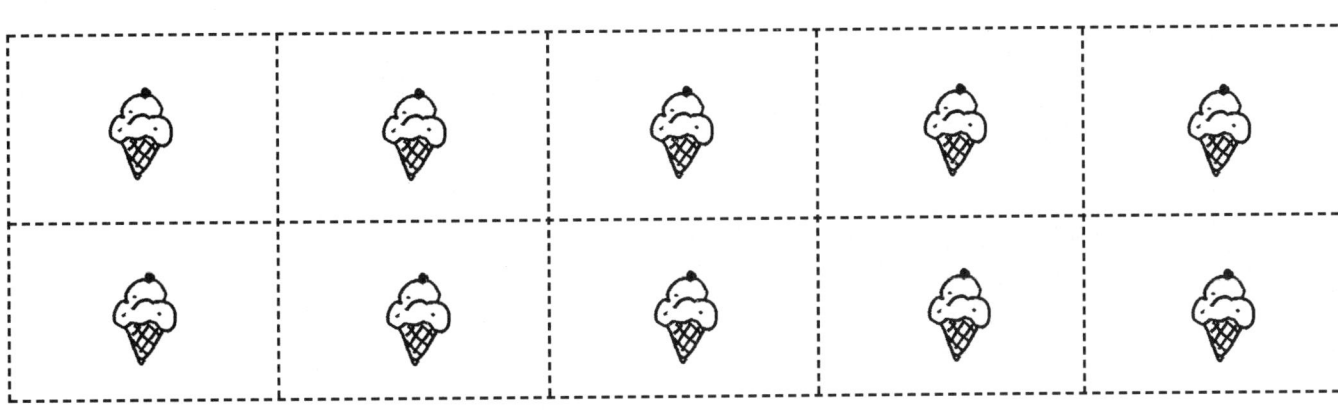

Data Analysis and Probability—graphs   Name _____

What Does It Mean?

Look at the bar graph and pictograph. Answer the questions.

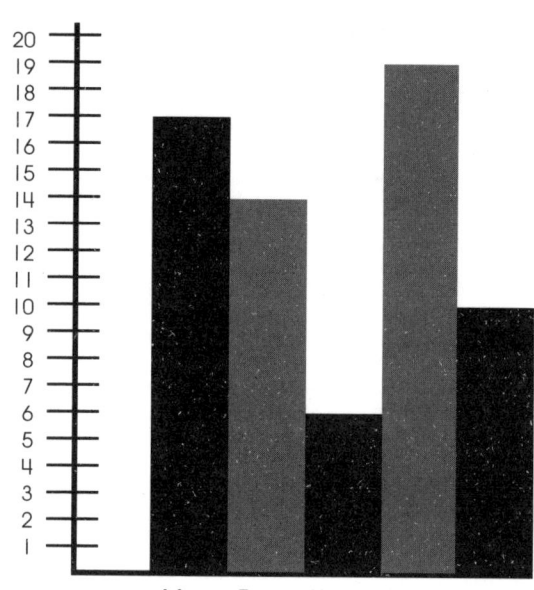
Number of Pages Read

1. How many more pages were read on Monday than on Tuesday?
   _____

2. What was the total number of pages read on Thursday and Friday?
   _____

3. How many more pages were read on Friday than on Wednesday? _____

4. How many pages were read on Thursday? _____

5. How many nonstop flights were there on Monday and Tuesday?
   _____

6. How many more nonstop flights were there on Wednesday than on Thursday?
   _____

7. How many nonstop flights were there on Tuesday? _____

8. What was the total number of nonstop flights all week?
   _____

Nonstop Flights from the Airport

Monday	✈✈
Tuesday	✈✈✈✈
Wednesday	✈✈✈
Thursday	✈✈
Friday	✈✈✈✈✈✈

✈ = 2 flights

© McGraw-Hill Children's Publishing   IF87125 Standards-Based Math

✓ Data Analysis and Probability—sort and classify   Name _____

Cut and Paste Facts

Look at the graph on this page. Write the number of rainy days on the line at the end of each box. Then complete the activity on page 110.

Rainy Days in March

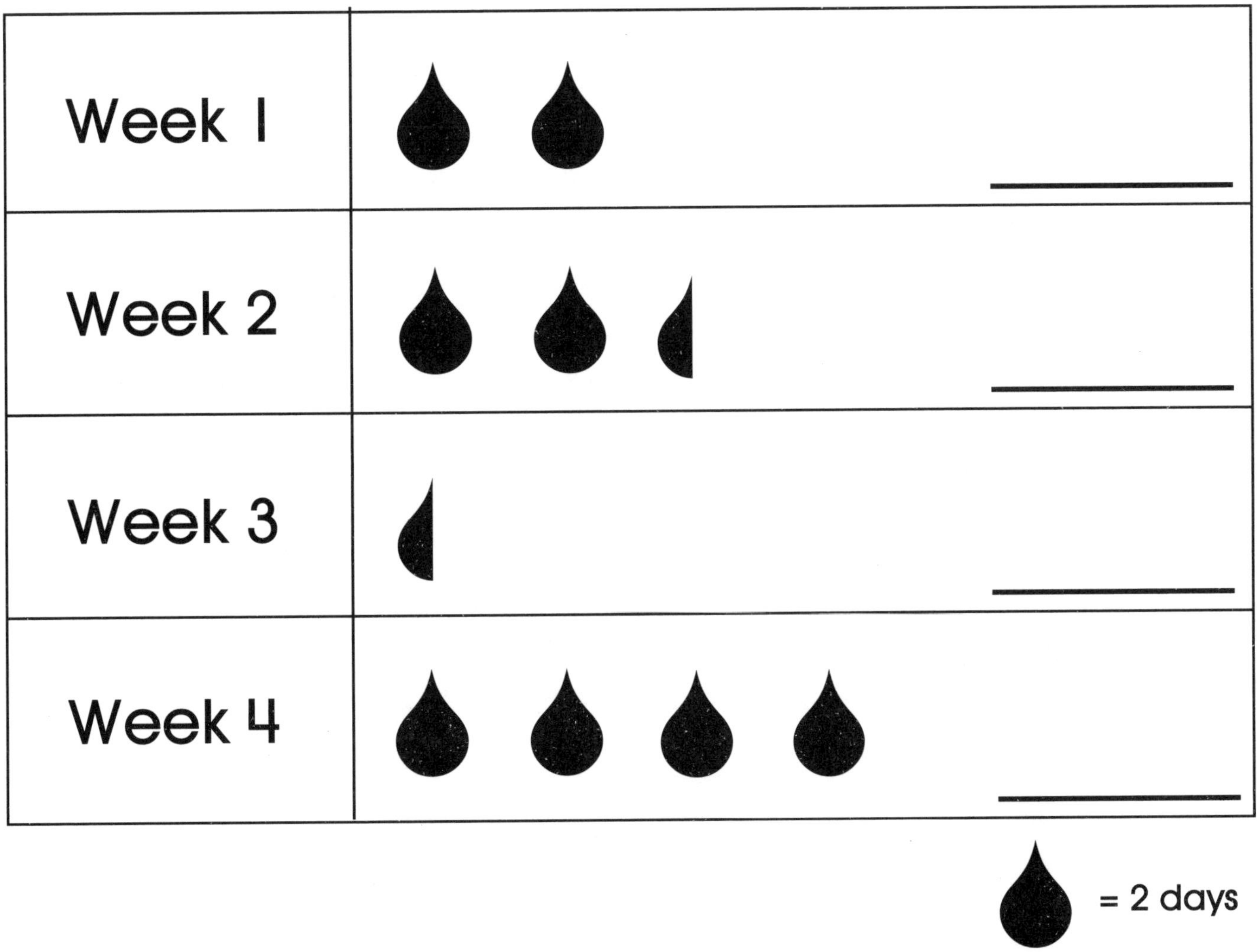

Week 1	🌢 🌢 _____
Week 2	🌢 🌢 🌗 _____
Week 3	🌗 _____
Week 4	🌢 🌢 🌢 🌢 _____

🌢 = 2 days

© McGraw-Hill Children's Publishing   IF87125 *Standards-Based Math*

Cut and Paste Facts, cont.

Name _____

Cut out the six facts below. Look at what you wrote on your graph on page 109. Glue the true statements below in the box labeled **True**. Glue the false statements in the box labeled **False**.

There were more rainy days in the last week than the first week.	There were 4 more rainy days in the third week than the first week.	There were 5 rainy days in the second week.
There were 15 rainy days in March.	There were 2 more rainy days in the last week than the first week.	There were 16 rainy days in March.

True	False

☑ Data Analysis and Probability—compiling data   Name _____

Important Information

Look at the chart. Fill in the pictograph and bar graph so that they show the same information as the chart. Don't forget titles.

Weekly Time Spent at the Computer Lab

Grade Level	Hours
First	3
Second	4
Third	4
Fourth	2
Fifth	4

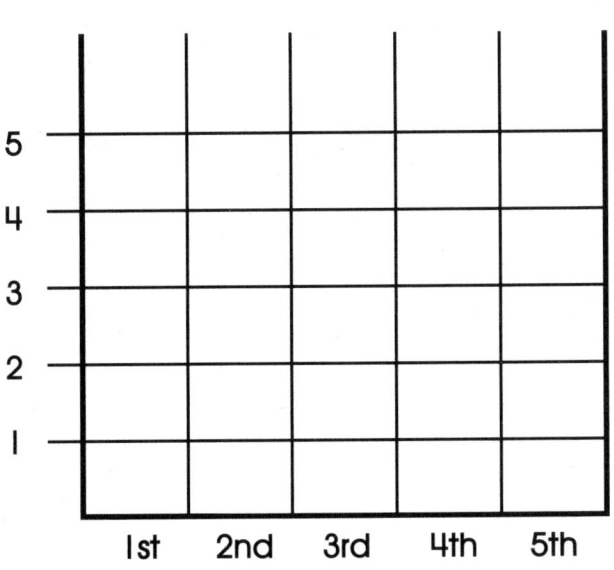

Bar Graph

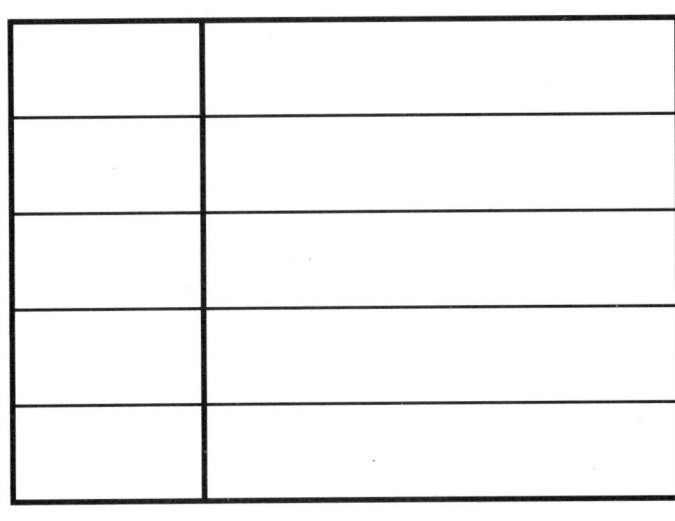

= 1 hour

Pictograph

© McGraw-Hill Children's Publishing   IF87125 Standards-Based Math

☑ Data Analysis and Probability—graphs   Name _____

Cold Weather Graphs

Answer the questions about the graph below.

1. How many more inches of snow fell during January than February?
   _____

2. How much snow fell between November and March?
   _____

3. How many inches of snow fell during November and December?
   _____

4. How much more snow fell during December than during November?
   _____

5. What was the difference in snowfall between December and March?
   _____

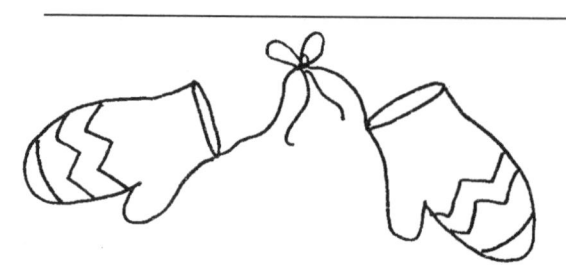

© McGraw-Hill Children's Publishing   IF87125 Standards-Based Math

Data Analysis and Probability—data analysis   Name _____

Money Graphs

Look at the graph. Answer the questions.

Sarah's Dog-Walking Earnings

May	¢ ¢ ¢ ¢ ¢ ¢ ¢ ¢ ¢
June	¢ ¢ ¢ ¢ ¢ ¢ ¢ ¢ ¢ ¢ ¢ ¢
July	¢ ¢ ¢ ¢ ¢ ¢ ¢ ¢ ¢ ¢ ¢ ¢ ¢ ¢
August	¢ ¢ ¢ ¢ ¢ ¢ ¢ ¢ ¢ ¢

¢ = 50¢

1. How much money did Sarah earn walking dogs in May?
   _____

2. How much money did Sarah earn during July and August?
   _____

3. How much more money did Sarah earn walking dogs in July than June?
   _____

4. During which month did Sarah earn the most money?
   _____

5. How much money did Sarah earn during all four months?
   _____

Now, create a pictograph of your own using the chart and the key.

Child	Money Spent
Hannah	$4.00
Andrea	$5.00
Peter	$8.00
Matt	$7.00

September Lunch Money

= $2.00

© McGraw-Hill Children's Publishing

Data Analysis and Probability—data analysis   Name _____

A Day at the Fair

Christine goes to the fair with $21.00. She spends all of her money. The pie graph below shows how Christine spent her money. Study the pie graph and then answer the questions.

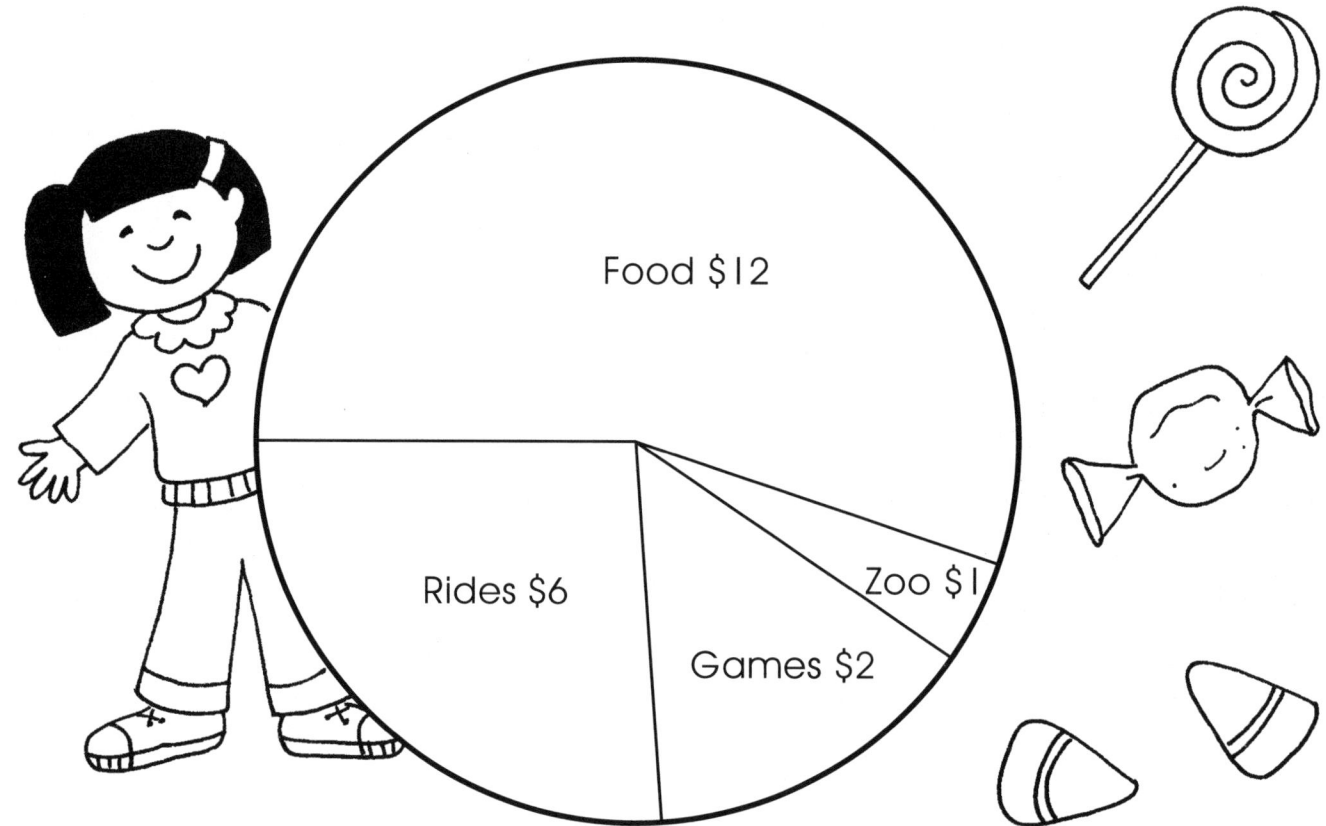

1. How did Christine spend most of her money?
   _____

2. How much money did Christine spend going to the petting zoo?
   _____

3. How much money did Christine spend on rides?
   _____

4. How much more money did Christine spend on eating than on playing games?
   _____

5. How much money did Christine spend playing games and going to the petting zoo?
   _____

Data Analysis and Probability—representing data   Name _____

Three Cheers for the Team!

Fill in the chart so that it shows the same facts as the pictograph.

Number of Goals Scored by the Lightning Bolts

September	⚽⚽
October	⚽◐
November	⚽⚽⚽
December	⚽⚽⚽⚽

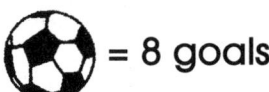 = 8 goals

Number of Goals Scored by the Lightning Bolts

Month	No. Goals
September	
October	
November	
December	

© McGraw-Hill Children's Publishing

Answer Key

Mix and Match Socks ....Page 6
1. 11
2. 7
3. 8
4. 9
5. 5
6. 16
7. 4
8. 14
9. 12

Ones and Tens ....Page 8
1. 4 tens, 5 ones
2. 1 ten, 4 ones
3. 0 tens, 9 ones
4. 1 ten, 0 ones
5. 6 tens, 6 ones
6. 1 ten, 7 ones
7. 0 tens, 8 ones
8. 4 tens, 1 one
9. 5 tens, 6 ones
10. 9 tens, 3 ones

Dot-to-dot reveals a monkey.

A House Full of Place Value ....Page 9
1. 3 hundreds, 0 tens, 0 ones
2. 2 hundreds, 1 ten, 1 one
3. 3 hundreds, 1 ten, 4 ones
4. 0 hundreds, 3 tens, 0 ones
5. 2 hundreds, 9 tens, 8 ones
6. 3 hundreds, 0 tens, 1 one

Mystery Numbers ....Page 10
1. 27
2. 10
3. 25
4. 103
5. 444
6. 8
7. 301
8. 12
9. 46
10. 798

Finding The "In Between" ....Page 11
1. 10
2. 20
3. 2
4. 15
5. 8
6. 23

Teddy Bears ....Page 12
1. match
2. 13
3. match
4. match
5. 11
6. 16
7. match
8. match

Understanding Fractions ....Page 13
Colored:
1. one carrot
2. one apple
3. one potato
4. one orange
5. one strawberry
6. one tomato
7. two cucumbers
8. one peach

Cupcake Fractions ....Page 14
1. $\frac{1}{4}$ of the cupcakes do not have hearts.
2. $\frac{1}{2}$ of the cupcakes have candles.
3. $\frac{1}{3}$ of the cupcakes have a cherry.

The Whole Team ....Page 15
1. one colored
2. one with a hat
3. one player with 12
4. one brown bat
5. two players with black hair
6. one girl with ribbon

Cut and Paste Fractions ....Page 16
1. $\frac{1}{4}$
2. $\frac{1}{3}$
3. $\frac{1}{5}$
4. $\frac{1}{2}$
5. $\frac{1}{2}$
6. $\frac{1}{10}$
7. $\frac{1}{5}$
8. $\frac{1}{4}$
9. $\frac{1}{3}$
10. $\frac{1}{2}$

Show What You Know! .... Page 17
1. 1 block green
2. 1 block purple
3. 1 block orange
4. 1 block blue
5. 1 block yellow
6. $\frac{1}{10}$
7. $\frac{1}{2}$
8. $\frac{1}{3}$
9. $\frac{1}{4}$
10. $\frac{1}{5}$

Who Is Bigger? .... Page 18
Colored: 10, 14, 19, 11, 18, 21, 12, 15, 13, 20, 16

Hungry Hippos .... Page 19
1. >; pink
2. <
3. <
4. >; pink
5. <
6. >; pink
7. <
8. <
9. >; pink
10. =

Compute with Caroline .... Page 20
1. 7—blue
2. 10—orange
3. 15—purple
4. 6—blue
5. 17—purple
6. 9—blue
7. 15—purple
8. 12—orange
9. 5—blue
10. 14—purple

Rover's Bones .... Page 21
1. 11
2. 11
3. 6
4. 14
5. 19
6. 13
7. 12
8. 16

Hot Air Balloons .... Page 22
1. 18; #9
2. 21; #7
3. 17; #10
4. 19; #8
5. 16; #6
6. 16; #5
7. 21; #2
8. 19; #4
9. 18; #1
10. 17; #3

Adding It Up .... Page 23
Color: 9 + 8; 10 + 7; 12 + 5; 8 + 9; 4 + 13; 16 + 1; 17 + 0; 14 + 3

1. 15
2. 24
3. 7
4. 17
5. 16
6. 17
7. 15
8. 17
9. 12
10. 18
11. 17
12. 17
13. 17
14. 17

Delicious Word Problems .... Page 24
1. 16
2. 10
3. 19
4. 11
5. 13

The Answer Is... ....Page 25
1. 3—orange
2. 4—brown
3. 8—red
4. 5—green
5. 3—orange
6. 3—orange
7. 0—brown
8. 3—orange

Sliding Through Subtraction ....Page 26
1. 6—green
2. 1—green
3. 1—green
4. 7—yellow
5. 3—green
6. 8—yellow
7. 6—green
8. 11—yellow
9. 7—yellow
10. 1—green

Bouncing Balls ....Page 27
1. 3
2. 4
3. 17
4. 3
5. 8
6. 8
7. 26
8. 8

Subtraction Practice ....Page 28
1. 18—pink
2. 14—pink
3. 7—orange
4. 2—orange
5. 7—orange
6. 6—orange
7. 6—orange
8. 13—pink
9. 2—orange
10. 17—pink
11. 15—pink
12. 7—orange
13. 6—orange
14. 9—orange
15. 3—orange

A Bunch of Bubbles ....Page 29
1. 21
2. 10
3. 7
4. 4
5. 14
6. 31
7. 11
8. 14
9. 11
10. 1
11. 12
12. 15

Bubbles colored: 19, 10, 7, 4, 14, 25, 11, 14, 9, 1, 8, 15

Worldly Word Problems ....Page 30
1. They have 11 pieces of gum left.
2. Janet catches 5 more fish.
3. There were 7 people not wearing hats.
4. Jack spends $2.00 more.
5. 7 taxis are not yellow.

Coins ....Page 31
1. 75¢ or $.75
2. 80¢ or $.80
3. 28¢ or $.28
4. $1.05
5. 43¢ or $.43
6. 60¢ or $.60

Maggie's Wallet..........................Page 32
1. 97¢; pencil
2. 19¢; eraser
3. 75¢; doughnuts
4. $1.11; ribbon
5. $1.55; ice cream cone

A Day at the Mall ....Page 33
1. 1 dime, 1 nickel
2. 3 quarters, 1 dime, 1 nickel, 1 penny
3. 5 quarters
4. 1 quarter, 2 dimes, 4 pennies
5. 3 quarters, 1 dime, 4 pennies

Louie's Loose Change ....Page 34
Monday: 72¢ or $.72
Tuesday: $1.65
Wednesday: 70¢ or $.70
Thursday: $1.01
Friday: $1.05
Saturday: 40¢ or $.40
Sunday: 25¢ or $.25

Coloring Coins ....Page 35
1. 55¢ or $.55—green
2. 34¢ or $.34—blue
3. 75¢ or $.75—yellow
4. $1.35—red
5. 48¢ or $.48—orange
6. $1.18—brown

Solve with Words ....Page 36
1. 70 bananas   2. 2 hours
3. 36 mice   4. 70 boys
5. 305 cards   6. 16 shade plants

Find the Sign ....Page 37
1. +; 17
2. –; 4
3. +; 13
4. –; 7
5. +; 11
6. –; 4
7. +; 13

Spending and Earning Money ....Page 38
1. +; $11.00
2. +; $27.00
3. –; $2.00
4. –; $2.00
5. +; $18.00

Practice, Practice ....Page 39
1. 104   2. 51
3. 92   4. 110
5. 104   6. 34
7. 110   8. 101
9. 90   10. 130
11. 81   12. 102
13. 127   14. 64
15. 150   16. 160

Mousetrap ....Page 40
1. 58   2. 50
3. 62   4. 103
5. 155   6. 191
7. 72   8. 165
9. 164   10. 63
11. 93   12. 121

Ice Cream Scoops ....Page 41
1. 165—blue   2. 103
3. 110   4. 162—blue
5. 111   6. 80
7. 161—blue   8. 102
9. 82   10. 190—blue
11. 57   12. 111

Adding Along ....Page 42
1. 104   2. 101
3. 92   4. 80
5. 84   6. 92
7. 104   8. 100
9. 101   10. 92
11. 112   12. 100
13. 106   14. 92
15. 100

Hats Off to You! ....Page 43
1. 35   2. 59
3. 37   4. 18
5. 76   6. 28
7. 29   8. 19

Subtraction Maze ....Page 44
1. 49   2. 17
3. 15   4. 31
5. 69   6. 29
7. 9   8. 79
9. 19   10. 38
11. 37   12. 39
13. 58   14. 69
15. 49

Problems colored: 1, 5, 8, 12, 13, 14, 15

A Piece of Cake! ....Page 45
1. 67 2. 59
3. 18 4. 35
5. 59 6. 8
7. 19 8. 32
9. 34 10. 17
11. 1 12. 17

Problems colored: 1, 2, 4, 5, 8, 9

Nutty Solutions ....Page 46
1. 52 2. 6
3. 37 4. 16
5. 8   6. 9
7. 19 8. 32
9. 34 10. 18

A Cool Drink ....Page 47
1. 37—red
2. 61—purple
3. 23—blue
4. 16—green
5. 26—blue
6. 33—red
7. 33—red
8. 46—yellow
9. 69—purple
10. 9—green
11. 34—red
12. 53—orange

Curtains Up on Addition ....Page 48
1. 394—blue
2. 506—red
3. 459—red
4. 332—blue
5. 676—red
6. 564—red
7. 342—blue
8. 519—red

You're Bright ....Page 49
1. 563—yellow
2. 971—orange
3. 1,328—orange
4. 828—orange
5. 946—orange
6. 516—yellow
7. 662—orange
8. 1,332—orange
9. 474—yellow
10. 676—orange
11. 762—orange
12. 855—orange

Sweet Treats ....Page 50
1. 510   2. 1,214—color
3. 815—color 4. 924—color
5. 715   6. 926—color
7. 790   8. 650

Up and Away ....Page 51
1. 1,035
2. 1,182
3. 1,010
4. 1,737
5. 1,276
6. 1,203
7. 776

Colored balloons: 1,035, 1,182, 1,010, 1,737, 1,276, 1,203, 776

Big Number on Top! ....Page 52
1. 118 2. 102
3. 159 4. 148
5. 219 6. 236
7. 869 8. 3

A Letter Maze .... Page 53
1. 175
2. 50
3. 31
4. 144
5. 484
6. 150
7. 98
8. 89
9. 145
10. 149

Problems Colored: 1, 4, 6, 9, 10

Tempting Snacks .... Page 54
1. 221
2. 65
3. 122
4. 426
5. 388
6. 113
7. 452
8. 148
9. 448
10. 157
11. 164
12. 169

Problems Circled: 4, 5, 7, 9

Secret Code .... Page 55
1. 99
2. 365
3. 455
4. 450
5. 135
6. 499
7. 290
8. 479
9. 177
10. 399
11. 230
12. 325
13. 299
14. 215

To prove he was not chicken

Groups .... Page 56
1. 4
2. 3
3. 8
4. 15
5. 1
6. 12
7. 12

Cookies, Cookies, Cookies! .... Page 57
1. 9
2. 15
3. 7
4. 12
5. 12
6. 15

Sharing with Friends .... Page 58
1. 2
2. 2
3. 4
4. 1

Birthday Party Guests .... Page 59
1. 1
2. 3
3. 2
4. 1
5. 4

Small to Large .... Page 60
1. 1, 2, 3
2. 3, 1, 2
3. 3, 2, 1
4. Drawings reflect directions.
5. Drawings reflect directions.

Sort It Out .... Page 61
4 sides: ruler, book, computer disk, picture frame, paper
Cylinder: pencil, cup, barrel, jar, flute

Skipping Stones .... Page 62
1. 9; 18
2. 10; 30; 60
3. 4; 8; 12
4. 25; 35; 40
5. 200; 400; 600
6. 24; 42

Number Patterns .... Page 63
1. 4
2. 6
3. 8
4. 25
5. 5
6. 18
7. 14
8. 12
9. 32
10. 3

Square, Square, Circle ....Page 64
1. triangle
2. square
3. pentagon
4. square; square
5. pentagon
6. octagon; triangle
7. circle
8. circle

Sizable Patterns ....Page 65
1. smiley face
2. shamrock
3. circle
4. pen
5. flower
6. chair
7. bird
8. sphere

Pasting Patterns ....Page 66
1. daisy
2. bone
3. car
4. square
5. mop

Candy Store Patterns ....Page 67
1. hard candy
2. candy cane
3. gumdrop
4. gumdrop
5. hard candy with stripes
6. chocolate bar
7. gumdrop
8. hard candy with stripes

It's the Opoosite ....Page 68
1. 7; 7
2. 7; 7
3. 6; 6
4. 17; 17

6. 3
   $\underline{+10}$
   13

7. 6
   $\underline{+7}$
   13

8. 6
   $\underline{+2}$
   8

Growing Up ....Page 69
1. taller
2. longer
3. shorter
4. bigger
5. shorter

Solving Symbols ....Page 70
1. 7   2. 12
3. 16   4. 14
5. 10   6. 10
7. 17   8. 13

Flat Shapes ....Page 71
1. 3   2. 4
3. 4   4. 5
5. 5   6. octagon
7. circle   8. hexagon
9. rectangle 10. pentagon

Flying High..................................Page 72
32 triangles—black
15 circles—orange
1 square—red
5 rectangles—green
1 pentagon—yellow
2 hexagons—purple
1 octagon—blue

Discovering Details ....Page 73

1. 11
2. 5
3. 29
4. 19
5. 4
6. 1
7. 1

Three-Dimensional Shapes ....Page 74
7. sphere
8. pyramid
9. cube
10. cone
11. cylinder
12. prism

In the Air ....Page 75

1. 25
2. 9
3. 0
4. 2
5. 1
6. 2
7. 1
8. 0
9. 2
10. 1
11. 1
12. 0
13. 1

Shapes Take Off ....Page 76

1.
2.
3.
4.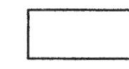
5.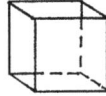
6.
7.
8.
9.
10.

Who Am I? ....Page 77
1. sphere
2. rectangle
3. cube
4. circle
5. cylinder

Symmetry ....Page 78
1. symmetrical
2. not symmetrical
3. not symmetrical
4. symmetrical
5. not symmetrical
6. symmetrical
7. not symmetrical
8. symmetrical

Find My Match ....Page 79
1.
2.
3.
4.
5.
6.
7.
8.

Simply Symmetrical ....Page 80

Mirror, Mirror, on the Wall ....Page 81
1. $\frac{1}{2}$ lamp
2. $\frac{1}{2}$ window
3. $\frac{1}{2}$ bow
4. $\frac{1}{2}$ hexagon
5. $\frac{1}{2}$ rug
6. $\frac{1}{2}$ M
7. $\frac{1}{2}$ circle
8. $\frac{1}{2}$ square

Lines of Symmetry ....Page 82
Correct: 1, 4, 6, 7

9. green
10. purple
11. purple
12. purple
13. green

Double Take ....Page 83
1. $\frac{1}{2}$ skirt
2. $\frac{1}{2}$ shoe
3. $\frac{1}{2}$ shirt
4. $\frac{1}{2}$ ribbon
5. $\frac{1}{2}$ shirt w/ button
6. $\frac{1}{2}$ pants

Shapes Around Us ....Page 84
1. pentagon
2. rectangle
3. octagon
4. sphere

5. triangle
6. hexagon
7. cone
8. cube
9.-13. Drawings will vary.

Symmetry Review..........................Page 85
1. $\frac{1}{2}$ heart
2. $\frac{1}{2}$ rectangle
3. $\frac{1}{2}$ circle
4. $\frac{1}{2}$ paper
5. $\frac{1}{2}$ cylinder
6. $\frac{1}{2}$ flower
7. $\frac{1}{2}$ triangle
8. $\frac{1}{2}$ bell

Counting Caterpillars ....Page 86
1. 3—yellow
2. 1—green
3. 4—yellow
4. 5—orange
5. 2—green
6. 6—orange
7. 8—red
8. 7—red

Paper Clip It ....Page 87
Using a standard 1 1/4 inch paper clip:
1. 2
2. 1
3. 1
4. 1
5. 5
6. $\frac{1}{2}$

Looking for Length ....Page 88
1. $2\frac{1}{2}$ inches
2. 2 inches
3. 5 inches
4. 6 inches
5. $4\frac{1}{2}$ inches

Taking the Measurement ....Page 89
1. 2 inches—green
2. $1\frac{1}{2}$ inches—brown
3. $2\frac{1}{2}$ inches—green
4. $\frac{1}{2}$ inch—brown
5. 1 inch—brown

Traveling with Squirrel ....Page 90
1. 5 meters
2. 18 meters
3. 11 meters
4. 4 meters
5. 29 meters

Munching ....Page 91
1. 4 cm – 2 cm = 2 cm
2. 7 cm – 3 cm = 4 cm
3. 6 cm – 2 cm = 4 cm
4. 5 cm – 2 cm = 3 cm
5. 9 cm – 5 cm = 4 cm
6. 3 cm – 1 cm = 2 cm

What Is the Wacky Weight?........Page 92
1. 1 pound—red
2. 3 pounds—red
3. 4 pounds—red
4. 1 pound—red
5. 5 pounds—blue
6. 1 pound—red
7. 4 pounds—red
8. 5 pounds—blue

Blue: bicycle, Lyle
Red: ball, pencil, kite, glasses, football, shirt, shoes

Basket Weight...............................Page 93
1. 3 kg
2. 4 kg
3. 2 kg
4. 5 kg
5. 1 kg
6. 6 kg

All About Area ....Page 94
1. 10 square units
2. 25 square units
3. 12 square units
4. 8 square units
5. 18 square units
6. 20 square units
7. 22 square units
8. 16 square units

Measurement Review ....Page 95
1. 3 inches
2. 2 inches
3. $4\frac{1}{2}$ inches
4. 1 pound
5. 11 pounds
6. 28 square units
7. 14 square units

At the Race ....Page 96
1. 1 quart
2. 1 quart
3. 8 cups
4. 4 pints

Telling Time ....Page 97
You better go catch it!

The Zookeeper's Watch ....Page 98

Don't Be Late! ....Page 99
1. 8:00—bath
2. 11:00—soccer
3. 5:00—football
4. 10:30—shop
5. 1:00—read
6. 3:30—party

Watching the Time ....Page 100

Time Puzzlers ....Page 101
1. 8:30; jogging
2. 10:15; Brenda
3. 11:00; storytime
4. 3:30; tennis
5. 4:45; grocery
6. 9:00; bed

Vacation Time! ....Page 102
1. 14 days
2. 2 months
3. 2 months
4. 3 days
5. 2 months
6. one-quarter

Taking Notes ....Page 103

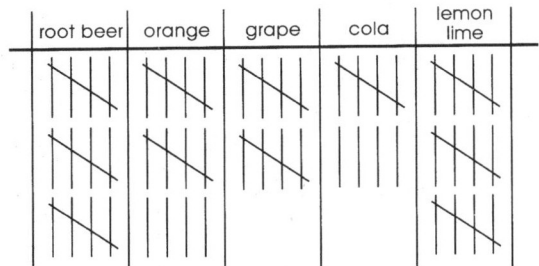

root beer and lemon lime

On the Chart ....Page 104

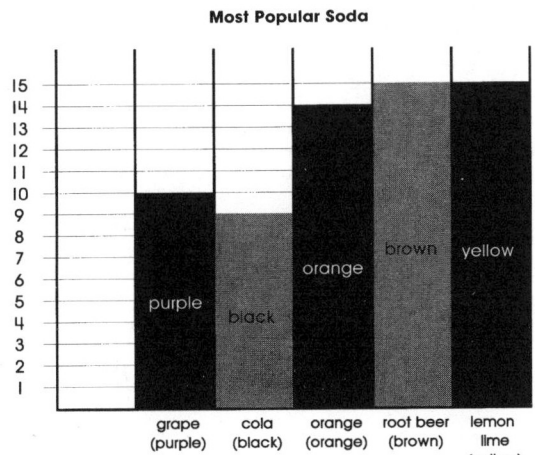

1. root beer and lemon lime
2. orange
3. cola

Pizza Pictograph ....Page 105
1. 16
2. 24
3. 4
4. 36
5. sausage

Guessable Graphs ....Page 106
1. Favorite Colors
2. Favorite Ice Cream Flavors
3. Weekly Telephone Calls
4. Number of Books Read by Second Graders

Matching the Missing ....Page 107

Chocolate (12)	
Vanilla (9)	
Strawberry (18)	
Cookie Dough (15)	

1. 3   2. 27
3. 54

What Does It Mean? ....Page 108
1. 3   2. 29
3. 4   4. 19
5. 12   6. 1
7. 8   8. 32

Cut and Paste Facts .... Page 109–110
True:
There were more rainy days in the last week than the first week.
There were 5 rainy days in the second week.
There were 4 more rainy days in the third week than the first week.
There were 16 rainy days in March.
False:
There were 15 rainy days in March.
There were 2 more rainy days in the last week than the first week.

Important Information .... Page 111

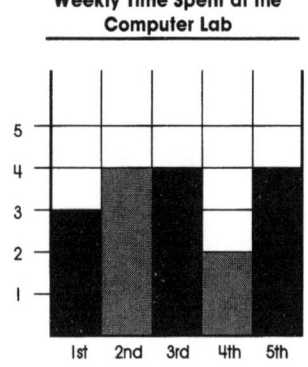

Bar Graph   Pictograph

Cold Weather Graphs .... Page 112
1. 12 inches
2. 110 inches
3. 46 inches
4. 10 inches
5. 16 inches

Money Graphs .... Page 113
1. $4.25
2. $12.75
3. $1.25
4. July
5. $23.00

A Day at the Fair .... Page 114
1. food
2. $1.00
3. $6.00
4. $10.00
5. $3.00

Three Cheers for the Team .... Page 115

Number of Goals Scored by the Lightning Bolts

Month	No. Goals
September	16
October	12
November	24
December	32

© McGraw-Hill Children's Publishing — IF87125 Standards-Based Math